Practical Variable Speed Drives and Power Electronics

Titles in the series

Practical Cleanrooms: Technologies and Facilities (David Conway)

Practical Data Acquisition for Instrumentation and Control Systems (John Park, Steve Mackay)

Practical Data Communications for Instrumentation and Control (John Park, Steve Mackay, Edwin Wright)

Practical Digital Signal Processing for Engineers and Technicians (Edmund Lai)

Practical Electrical Network Automation and Communication Systems (Cobus Strauss)

Practical Embedded Controllers (John Park)

Practical Fiber Optics (David Bailey, Edwin Wright)

Practical Industrial Data Networks: Design, Installation and Troubleshooting (Steve Mackay, Edwin Wright, John Park, Deon Reynders)

Practical Industrial Safety, Risk Assessment and Shutdown Systems (Dave Macdonald)

Practical Modern SCADA Protocols: DNP3, 60870.5 and Related Systems (Gordon Clarke, Deon Reynders)

Practical Radio Engineering and Telemetry for Industry (David Bailey)

Practical SCADA for Industry (David Bailey, Edwin Wright)

Practical TCP/IP and Ethernet Networking (Deon Reynders, Edwin Wright)

Practical Variable Speed Drives and Power Electronics (Malcolm Barnes)

Practical Variable Speed Drives and Power Electronics

Malcolm Barnes CPEng, BSc(ElecEng), MSEE, Automated Control Systems, Perth, Australia

AMSTERDAM • BOSTON • HEIDELBERG • LONDON • NEW YORK • OXFORD
PARIS • SAN DIEGO • SAN FRANCISCO • SINGAPORE • SYDNEY • TOKYO

Newnes is an imprint of Elsevier

Newnes

Newnes
An imprint of Elsevier
Linacre House, Jordan Hill, Oxford OX2 8DP
200 Wheeler Road, Burlington, MA 01803

First published 2003

British Library Cataloguing in Publication Data
A catalogue record for this book is available from the British Library

ISBN   07506 58088

For information on all Newnes publications, visit
our website at www.newnespress.com

Typeset and Edited by Vivek Mehra, Mumbai, India

Transferred to Digital Print 2009
Printed and bound in Great Britain by
CPI Antony Rowe, Chippenham and Eastbourne

Contents

3   Power electronic converters   68

4   Electromagnetic compatibility (EMC)   114

7 Selection of AC converters 178

8 Installation and commissioning 209

Preface

The rapid adoption of automation techniques in industry has increased the requirement for better process control. This has resulted in many new applications for AC variable speed drives (VSDs) to control the speed and torque of driven machinery. Variable speed drives (VSDs) are also used to meet particular starting and stopping requirements.

The variable speed drives book promotes a sound understanding of how VSDs work and how to correctly select, install, commission and maintain them. There is also detailed coverage of many typical applications in process control and materials handling such as those for pumping, ventilation, conveyers and hoists.

This book will benefit anyone associated with the use of VSDs in the industrial or automation environment. This book will also benefit those working in system design as well as site commissioning, maintenance and troubleshooting.

Although a basic understanding of electrical engineering principles is essential, even those with a superficial knowledge of VSDs will substantially benefit from this book.

In particular, if you work in any of the following areas, you will benefit from this book:

- Consulting electrical engineers
- Plant engineers and instrument technicians
- Operations technicians
- Electrical maintenance technicians and supervisors
- Instrumentation and control system engineers
- Process control engineers
- Mechanical engineers

We would hope that you will learn the following from this book:

- The principles of AC variable speed drives for industrial speed control
- The essentials of squirrel cage induction motors
- The latest developments in power electronic converters used for VSDs
- How to select the correct AC variable speed drive for industrial applications
- How to identify faults on VSDs and how to rectify them
- The key issues about flux vector control and how it can be used in drive applications
- The main concepts in interfacing the control circuits of VSDs with PLCs/DCSs using serial data communications

The structure of the book is as follows.

Chapter 1: Introduction. A review of the fundamentals in variable speed drives including motion concepts, torque speed curves, types of variable speed drives, mechanical variable speed drive methods and electrical variable speed drive methods.

Chapter 2: 3-phase AC induction motors. These versatile and robust devices are the prime movers for the vast majority of machines. This chapter covers the basic construction, electrical and mechanical performance, motor acceleration, AC induction generator performance, efficiency of electric motors, rating of AC induction motors, duty cycles, cooling and ventilation, degree of protection of motor enclosures, methods of starting and motor selection.

Chapter 3: Power electronic converters. This chapter deals with the active components (e.g. diodes, thyristors, transistors) and passive components (e.g. resistors, chokes, capacitors) used in power electronic circuits and converters.

Chapter 4: Electromagnetic compatibility (EMC). Interference in circuits refers to the presence of unwanted voltages or currents in electrical equipment, which can damage the equipment or degrade its performance. The impact of variable speed drives can be severe and this chapter examines what causes interference and how to minimize its impact.

Chapter 5: Protection of AC converters and motors. The protection of AC variable speed drives includes the protection of the AC converter and the electric motor. The main methods of protection are examined.

Chapter 6: Control systems for AC variable speed drives. The overall control system can be divided into four main areas of the inverter control system, speed feedback and control system, current feedback and control system and the external interface.

Chapter 7: Selection of AC converters. Although manufacturers' catalogs try to make it as easy as possible, there are many variables associated with the selection and rating of the optimum electric motor and AC converter for a VSD application. This chapter covers many of the principles for the correct selection for AC variable speed drives, which use pwm-type variable voltage variable frequency (VVVF) converters to control the speed of standard AC squirrel cage induction motors.

Chapter 8: Installation and commissioning. The main issues here of general installation and environmental requirements, power supply and earthing requirements, start/stop of AC drives, installing AC converters into metal enclosures, control wiring and commissioning variable speed drives.

Chapter 9: Special topics and new developments. Typical topics of soft-switching and the matrix converter are examined here.

1

Introduction

1.1   The need for variable speed drives

There are many and diverse reasons for using variable speed drives. Some applications, such as paper making machines, cannot run without them while others, such as centrifugal pumps, can benefit from energy savings.

In general, variable speed drives are used to:

- **Match the speed of a drive to the process requirements**
- **Match the torque of a drive to the process requirements**
- **Save energy and improve efficiency**

The needs for speed and torque control are usually fairly obvious. Modern electrical VSDs can be used to accurately maintain the speed of a driven machine to within ±0.1%, independent of load, compared to the speed regulation possible with a conventional fixed speed squirrel cage induction motor, where the speed can vary by as much as 3% from no load to full load.

The benefits of energy savings are not always fully appreciated by many users. These savings are particularly apparent with centrifugal pumps and fans, where load torque increases as the square of the speed and power consumption as the cube of the speed. Substantial cost savings can be achieved in some applications.

An everyday example, which illustrates the benefits of variable speed control, is the motorcar. It has become such an integral part of our lives that we seldom think about the technology that it represents or that it is simply a variable speed platform. It is used here to illustrate how variable speed drives are used to improve the speed, torque and energy performance of a machine.

It is intuitively obvious that the speed of a motorcar must continuously be controlled by the driver (the operator) to match the traffic conditions on the road (the process). In a city, it is necessary to obey speed limits, avoid collisions and to start, accelerate, decelerate and stop when required. On the open road, the main objective is to get to a destination safely in the shortest time without exceeding the speed limit. The two main controls that are used to control the speed are the accelerator, which controls the driving torque, and the brake, which adjusts the load torque. A motorcar could not be safely operated in city traffic or on the open road without these two controls. The driver must continuously adjust the fuel input to the engine (the drive) to maintain a constant speed in spite of the changes in the load, such as an uphill, downhill or strong wind conditions. On other occasions he may have to use the brake to adjust the load and slow the vehicle down to standstill.

Another important issue for most drivers is the cost of fuel or the cost of energy consumption. The speed is controlled via the accelerator that controls the fuel input to the engine. By adjusting the accelerator position, the energy consumption is kept to a minimum and is matched to the speed and load conditions. Imagine the high fuel consumption of a vehicle using a fixed accelerator setting and controlling the speed by means of the brake position.

1.2 Fundamental principles

The following is a review of some of the fundamental principles associated with variable speed drive applications.

- **Forward direction**
  Forward direction refers to motion in one particular direction, which is chosen by the user or designer as being the forward direction. The Forward direction is designated as being positive (+ve). For example, the forward direction for a motorcar is intuitively obvious from the design of the vehicle. Conveyor belts and pumps also usually have a clearly identifiable forward direction.

- **Reverse direction**
  Reverse direction refers to motion in the opposite direction. The Reverse direction is designated as being negative (–ve). For example, the reverse direction for a motor car is occasionally used for special situations such as parking or un-parking the vehicle.

- **Force**
  Motion is the result of applying one or more forces to an object. Motion takes place in the direction in which the resultant force is applied. So force is a combination of both **magnitude and direction**. A Force can be +ve or –ve depending on the direction in which it is applied. A Force is said to be +ve if it is applied in the forward direction and –ve if it is applied in the reverse direction. In SI units, force is measured in *Newtons*.

- **Linear velocity (v) or speed (n)**
  Linear velocity is the measure of the linear distance that a moving object covers in a unit of time. It is the result of a linear force being applied to the object. In SI units, this is usually measured in *meters per second (m/sec)*. *Kilometers per hour (km/hr)* is also a common unit of measurement. For motion in the forward direction, velocity is designated Positive (+ve). For motion in the reverse direction, velocity is designated Negative (–ve).

- **Angular velocity (ω) or rotational speed (n)**
  Although a force is directional and results in linear motion, many industrial applications are based on rotary motion. The rotational force associated with rotating equipment is known as torque. Angular velocity is the result of the

application of torque and is the angular rotation that a moving object covers in a unit of time. In SI units, this is usually measured in radians per second (rad/sec) or revolutions per second (rev/sec). When working with rotating machines, these units are usually too small for practical use, so it is common to measure rotational speed in revolutions per minute (rev/min).

- **Torque**
  Torque is the product of the tangential force F, at the circumference of the wheel, and the radius r to the center of the wheel. In SI units, torque is measured in Newton-meters (Nm). A torque can be +ve or −ve depending on the direction in which it is applied. A torque is said to be +ve if it is applied in the forward direction of rotation and −ve if it is applied in the reverse direction of rotation.

Using the motorcar as an example, Figure 1.1 illustrates the relationship between direction, force, torque, linear speed and rotational speed. The petrol engine develops rotational torque and transfers this via the transmission and axles to the driving wheels, which convert torque (T) into a tangential force (F). No horizontal motion would take place unless a resultant force is exerted horizontally along the surface of the road to propel the vehicle in the forward direction. The higher the magnitude of this force, the faster the car accelerates. In this example, the motion is designated as being forward, so torque, speed, acceleration are all +ve.

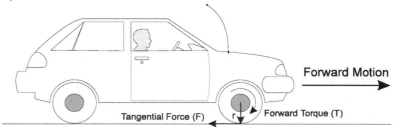

Torque (Nm) = Tangential Force (N) × Radius (m)

Figure 1.1:
The relationship between torque, force and radius

- **Linear acceleration (a)**
  Linear acceleration is the rate of change of linear velocity, usually in m/sec^2.

$$\text{Linear acceleration} \qquad a = \frac{\mathrm{d}v}{\mathrm{d}t} \quad \text{m/sec}^2$$

  – Linear acceleration is the increase in velocity in either direction

  – Linear deceleration or braking is the decrease in velocity in either direction

- **Rotational acceleration (a)**
  Rotational acceleration is the rate of change of rotational velocity, usually in rad/sec^2.

$$\text{Rotational acceleration} \qquad a = \frac{\mathrm{d}\omega}{\mathrm{d}t} \quad \text{rad/sec}^2$$

- Rotational acceleration is the increase in velocity in either direction
- Rotational deceleration or Braking is the decrease in **velocity** in either direction

In the example in Figure 1.2, a motorcar sets off from standstill and accelerates in the forward direction up to a velocity of 90 km/hr (25 m/sec) in a period of 10 sec.

In variable speed drive applications, this acceleration time is often called the *ramp-up time*. After traveling at 90 km/hr for a while, the brakes are applied and the car decelerates down to a velocity of 60 km/hr (16.7 m/sec) in 5 sec. In variable speed drive applications, this deceleration time is often called the *ramp-down time*.

FORWARD DIRECTION

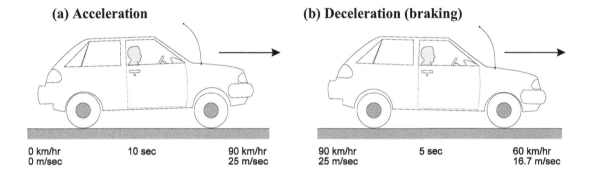

(a) Acceleration

0 km/hr   10 sec   90 km/hr
0 m/sec   25 m/sec

(b) Deceleration (braking)

90 km/hr   5 sec   60 km/hr
25 m/sec   16.7 m/sec

$$Acceleration = \frac{v_2 - v_1}{t} = \frac{25 - 0}{10} = +2.5$$

$$Acceleration = \frac{v_2 - v_1}{t} = \frac{16.7 - 25}{5} = -1.67 \text{ m/sec}^2$$

Figure 1.2:
Acceleration and deceleration (braking) in the forward direction

REVERSE DIRECTION

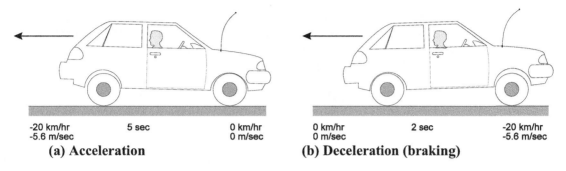

| (a) Acceleration | (b) Deceleration (braking) |

Figure 1.3:
Acceleration and deceleration (braking) in the reverse direction

$$Acceleration = \frac{v_2 - v_1}{t} = \frac{-5.6 - 0}{5} = -1.12 \text{ m/sec}^2$$

$$Acceleration = \frac{v_2 - v_1}{t} = \frac{0 - (-5.6)}{2} = +2.8 \text{ m/sec}^2$$

From the example outlined in Figure 1.3, the acceleration time (ramp-up time) to 20 km/hr in the reverse direction is 5 secs. The braking period (ramp-down time) back to standstill is 2 sec.

There are some additional terms and formulae that are commonly used in association with variable speed drives and rotational motion.

- **Power**

  Power is the rate at which work is being done by a machine. In SI units, it is measured in watts. In practice, power is measured in kiloWatts (kW) or MegaWatts (MW) because watts are such a small unit of measurement.

  In rotating machines, power can be calculated as the product of torque and speed. Consequently, when a rotating machine such as a motor car is at standstill, the output power is zero. This does not mean that input power is zero! Even at standstill with the engine running, there are a number of power losses that manifest themselves as heat energy.

  Using SI units, power and torque are related by the following very useful formula, which is used extensively in VSD applications:

  $$Power \text{ (kW)} = \frac{Torque \text{ (Nm)} \times Speed \text{ (rev/min)}}{9550}$$

  Alternatively,

$$Torque\,(\text{Nm}) = \frac{9550 \times Power\,(\text{kW})}{Speed\,(\text{rev}/\text{min})}$$

- **Energy**
  Energy is the product of power and time and represents the rate at which work is done over a period of time. In SI units it is usually measured as kiloWatt-hours (kWh). In the example of the motorcar, the fuel consumed over a period of time represents the energy consumed.

$$Energy\,(\text{kWh}) = Power\,(\text{kW}) \times Time\,(\text{h})$$

- **Moment of Inertia**
  Moment of inertia is that property of a rotating object that resists change in rotational speed, either acceleration or deceleration. In SI units, moment of inertia is measured in kgm^2.

  This means that, to accelerate a rotating object from speed n_1 (rev/min) to speed n_2 (rev/min), an acceleration torque T_A (Nm) must be provided by the prime mover in addition to the mechanical load torque. The time t (sec) required to change from one speed to another will depend on the moment of inertia J (kgm^2) of the rotating system, comprising both the drive and the mechanical load. The acceleration torque will be:

$$T_A\,(\text{Nm}) = J\,(\text{kgm}^2) \times \frac{2\pi}{60} \times \frac{(n_2 - n_1)\,(\text{rev/m})}{t\,(\text{sec})}$$

  In applications where rotational motion is transformed into linear motion, for example on a crane or a conveyor, the rotational speed (n) can be converted to linear velocity (v) using the diameter (d) of the rotating drum as follows:

$$v\,(\text{m/sec}) = \pi\,d\,n\,(\text{rev/sec}) = \frac{\pi\,d\,n\,(\text{rev/min})}{60}$$

therefore

$$T_A\,(\text{Nm}) = J\,(\text{kgm}^2) \times \frac{2}{d} \times \frac{(v_2 - v_1)\,(\text{m/sec})}{t\,(\text{sec})}$$

From the above power, torque and energy formulae, there are four possible combinations of acceleration/braking in either the forward/reverse directions that can be applied to this type of linear motion. Therefore, the following conclusions can be drawn:

- **1st QUADRANT, torque is +ve and speed is +ve.**
  Power is positive in the sense that energy is transferred from the prime mover (engine) to the mechanical load (wheels).

This is the case of the machine driving in the forward direction.

- **2nd QUADRANT, torque is –ve and speed is +ve.**
  Power is negative in the sense that energy is transferred from the wheels back to the prime mover (engine). In the case of the motor car, this returned energy is wasted as heat. In some types of electrical drives this energy can be transferred back into the power supply system, called *regenerative braking*.

This is the case of the machine braking in the forward direction.

> - **3rd QUADRANT, torque is –ve and speed is –ve.**
>   Power is positive in the sense that energy is transferred from the prime mover (engine) to the mechanical load (wheels).

This is the case of the machine driving in the reverse direction.

> - **4th QUADRANT, If torque is +ve and speed is –ve.**
>   Power is negative in the sense that energy is transferred from the wheels back to the prime mover (engine). As above, in some types of electrical drives this power can be transferred back into the power supply system, called regenerative braking.

This is the case of the machine braking in the reverse direction.

These 4 quadrants are summarized in Figure 1.4.

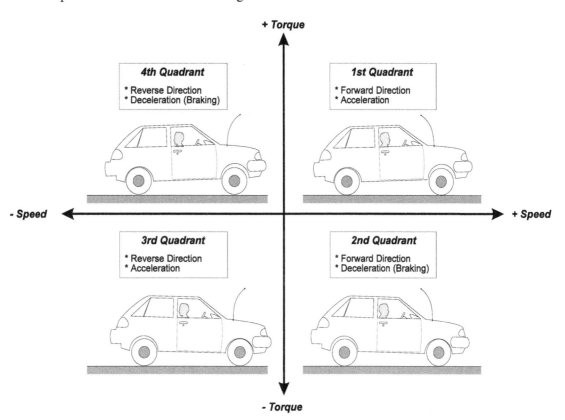

Figure 1.4:
The four quadrants of the torque-speed diagram for a motor car

1.3   Torque-speed curves for variable speed drives

In most variable speed drive applications torque, power, and speed are the most important parameters. Curves, which plot torque against speed on a graph, are often used to illustrate the performance of the VSD. The speed variable is usually plotted along one axis and the torque variable along the other axis. Sometimes, power is also plotted along the same axis as the torque. Since energy consumption is directly proportional to power, energy depends on the product of torque and speed. For example, in a motorcar, depressing the accelerator produces more torque that provides acceleration and results in more speed, but more energy is required and more fuel is consumed.

Again using the motorcar as an example of a variable speed drive, torque–speed curves can be used to compare two alternative methods of speed control and to illustrate the differences in energy consumption between the two strategies:

- **Speed controlled by using drive control**: adjusting the torque of the prime mover. In practice, this is done by adjusting the fuel supplied to the engine, using the accelerator for control, without using the brake. This is analogous to using an electric variable speed drive to control the flow of water through a centrifugal pump.
- **Speed controlled by using load control**: adjusting the overall torque of the load. In practice, this could be done by keeping a fixed accelerator setting and using the brakes for speed control. This is analogous to controlling the water flow through a centrifugal pump by throttling the fluid upstream of the pump to increase the head.

Using the motorcar as an example, the two solid curves in Figure 1.5 represent the drive torque output of the engine over the speed range for two fuel control conditions:

- High fuel position – accelerator full down
- Lower fuel position – accelerator partially down

The two dashed curves in the Figure 1.5 represent the load torque changes over the speed range for two mechanical load conditions. The mechanical load is mainly due to the wind resistance and road friction, with the restraining torque of the brakes added.

- Wind & friction plus brake ON – high load torque
- Wind & friction plus brake OFF – low load torque

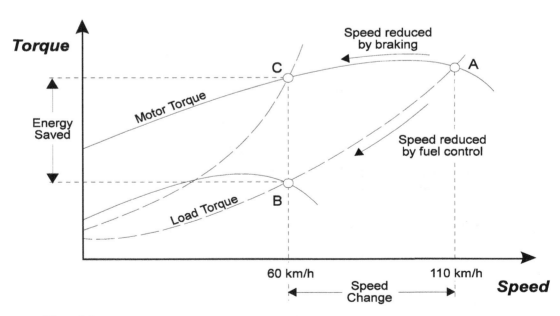

Figure 1.5:
Torque–speed curves for a motorcar

As with any drive application, a *stable speed* is achieved when the drive torque is equal to the load torque, where the drive torque curve intersects with the load torque curve. The following conclusions can be drawn from Figure 1.5 and also from personal experience driving a motor car:

- Fixed accelerator position, if load torque increases (uphill), speed drops
- Fixed accelerator position, if load torque decreases (downhill), speed increases
- Fixed load or brake position, if drive torque increases by increasing the fuel, speed increases (up to a limit)
- Fixed load or brake position, if drive torque decreases by reducing the fuel, speed decreases

As an example, assume that a motorcar is traveling on an open road at a stable speed with the brake off and accelerator partially depressed. The main load is the wind resistance and road friction. The engine torque curve and load torque curve cross at point A, to give a stable speed of 110 km/h. When the car enters the city limits, the driver needs to reduce speed to be within the 60 km/h speed limit. This can be achieved in one of the two ways listed above:

- Fuel input is reduced, speed decreases along the load-torque curve A–B. As the speed falls, the load torque reduces mainly due to the reduction of wind resistance. A new stable speed of 60 km/h is reached at a new intersection of the load–torque curve and the engine–torque curve at point B.
- The brake is applied with a fixed fuel input setting, speed decreases along the drive-torque curve A–C due to the increase in the load torque. A new stable speed is reached when the drive–torque curve intersects with the steeper load–torque curve at 60 km/h.

As mentioned previously, the power is proportional to Torque × Speed:

$$Power\,(\text{kW}) = \frac{Torque\,(\text{Nm}) \times Speed(\text{rev}/\min)}{9550}$$

$$Energy\,(\text{kWh}) = \frac{Torque\,(\text{Nm}) \times Speed\,(\text{rev}/\min) \times Time\,(\text{h})}{9550}$$

In the motor car example, what is the difference in energy consumption between the two different strategies at the new stable speed of 60 km/h? The drive speed control method is represented by Point B and the brake speed control method is represented by Point C. From above formula, the differences in energy consumption between points B and C are:

$$E_C - E_B = \frac{T_C\,60\,t}{9550} - \frac{T_B\,60\,t}{9550}$$

$$E_C - E_B = k\,(T_C - T_B)$$

The energy saved by using drive control is directly proportional to the difference in the load torque associated with the two strategies. This illustrates how speed control and energy savings can be achieved by using a variable speed drive, such as a petrol engine, in a motorcar. The added advantages of a variable speed drive strategy are the reduced wear on the transmission, brakes and other components.

The same basic principles apply to industrial variable speed drives, where the control of the speed of the prime mover can be used to match the process conditions. The control can be achieved manually by an operator. With the introduction of automation, speed control can be achieved automatically, by using a feedback controller which can be used to maintain a process variable at a preset level. Again referring to the motorcar example, automatic speed control can be achieved using the 'auto-cruise' controller to maintain a constant speed on the open road.

Another very common application of VSDs for energy savings is the speed control of a centrifugal pump to control fluid flow. Flow control is necessary in many industrial applications to meet the changing demands of a process. In pumping applications, Q–H curves are commonly used instead of torque–speed curves for selecting suitable pumping characteristics and they have many similarities. Figure 1.6 shows a typical set of Q–H curves. Q represents the flow, usually measured in m^3/h and H represents the head, usually measured in meters. These show that when the pressure head increases on a centrifugal pump, the flow decreases and vice versa. In a similar way to the motor car example above, fluid flow through the pump can be controlled either by controlling the speed of the motor driving the pump or alternatively by closing an upstream control valve (throttling). Throttling increases the effective head on the pump that, from the Q–H curve, reduces the flow.

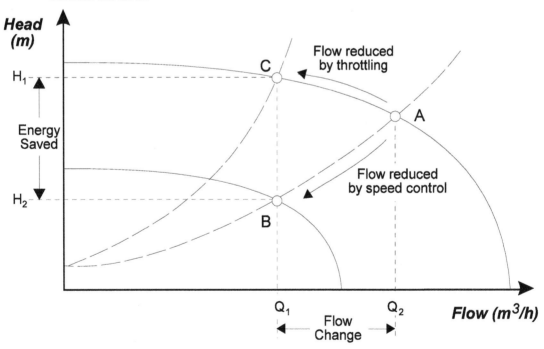

Figure 1.6:
Typical Q–H curves for a centrifugal pump

From Figure 1.6, the reduction of flow from Q_2 to Q_1 can be achieved by using one of the following two alternative strategies:

- **Drive speed control**, flow decreases along the curve A–B and to a point on another Q–H curve. As the speed falls, the pressure/head reduces mainly due to the reduction of friction in the pipes. A new stable flow of Q_1 m³/h is reached at point B and results in a head of H_2.
- **Throttle control**, an upstream valve is partially closed to restrict the flow. As the pressure/head is increased by the valve, the flow decreases along the curve A–C. The new stable flow of Q_1 m³/h is reached at point C and results in a head of H_1.

From the well-known pump formula, the power consumed by the pump is:

Pump Power (kW) = $k \times$ *Flow* (m³/h) $\times$ *Head* (m)

Pump Power (kW) = $k \times Q \times H$

Absorbed Energy (kWh) = $k \times Q \times H \times t$

$E_C - E_B = (kQ_1H_1t) - (kQ_1H_2t)$

$E_C - E_B = kQ_1 (H_1 - H_2)t$

$E_C - E_B = K (H_1 - H_2)$

With flow constant at Q_1, the energy saved by using drive speed control instead of throttle control is directly proportional to the difference in the head associated with the two strategies. The energy savings are therefore a function of the difference in the head between the point B and point C. The energy savings on large pumps can be quite substantial and these can readily be calculated from the data for the pump used in the application.

There are other advantages in using variable speed control for pump applications:

- Smooth starting, smooth acceleration/deceleration to reduce mechanical wear and water hammer.
- No current surges in the power supply system.
- Energy savings are possible. These are most significant with centrifugal loads such as pumps and fans because power/energy consumption increases/decreases with the cube of the speed.
- Speed can be controlled to match the needs of the application. This means that speed, flow or pressure can be accurately controlled in response to changes in process demand.
- Automatic control of the process variable is possible, for example to maintain a constant flow, constant pressure, etc. The speed control device can be linked to a process control computer such as a PLC or dcS.

1.4 Types of variable speed drives

The most common types of variable speed drives used today are summarized below:

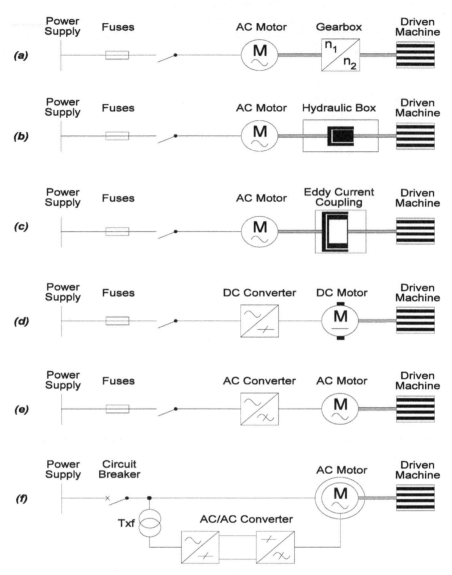

Figure 1.7:
Main types of variable speed drive for industrial applications

(a) Typical mechanical VSD with an AC motor as the prime mover;

(b) Typical hydraulic VSD with an AC motor as the prime mover;

(c) Typical electromagnetic coupling or Eddy Current coupling;

(d) Typical electrical VSD with a DC motor and DC voltage converter;

(e) Typical electrical VSD with an AC motor and AC frequency converter;

(f) Typical slip energy recovery system or static Kramer system;

Variable speed drives can be classified into three main categories, each with their own advantages and disadvantages:

- **Mechanical variable speed drives**
  - Belt and chain drives with adjustable diameter sheaves
  - Metallic friction drives

- **Hydraulic variable speed drives**
  - Hydrodynamic types
  - Hydrostatic types

- **Electrical variable speed drives**
  - Schrage motor (AC commutator motor)
  - Ward-Leonard system (AC motor – DC generator – DC motor)
  - Variable voltage DC converter with DC motor
  - Variable voltage variable frequency converter with AC motor
  - Slip control with wound rotor induction motor (slipring motor)
  - Cycloconverter with AC motor
  - Electromagnetic coupling or 'Eddy Current' coupling
  - Positioning drives (servo and stepper motors)

1.5 Mechanical variable speed drive methods

Historically, electrical VSDs, even DC drives, were complex and expensive and were only used for the most important or difficult applications. So mechanical devices were developed for insertion between a fixed speed electric drive motor and the shaft of the driven machine.

Mechanical variable speed drives are still favored by many engineers (mainly mechanical engineers!) for some applications mainly because of simplicity and low cost.

As listed above, there are basically 2 types of mechanical construction.

1.5.1 Belt and chain drives with adjustable diameter sheaves

The basic concept behind adjustable sheave drives is very similar to the gear changing arrangement used on many modern bicycles. The speed is varied by adjusting the ratio of the diameter of the drive pulley to the driven pulley.

For industrial applications, an example of a continuously adjustable ratio between the drive shaft and the driven shaft is shown in Figure 1.8. One or both pulleys can have an adjustable diameter. As the diameter of one pulley increases, the other decreases thus maintaining a nearly constant belt length. Using a V-type drive belt, this can be done by adjusting the distance between the tapered sheaves at the drive end, with the sheaves at the other end being spring loaded. A hand-wheel can be provided for manual control or a servo-motor can be fitted to drive the speed control screw for remote or automatic control. Ratios of between 2:1 and 6:1 are common, with some low power units capable of up to 16:1. When used with gear reducers, an extensive range of output speeds and gear ratios are possible. This type of drive usually comes as a totally enclosed modular unit with an AC motor fitted. On the chain version of this VSD, the chain is usually in the form of a wedge type roller chain, which can transfer power between the chain and the smooth surfaces of the tapered sheaves.

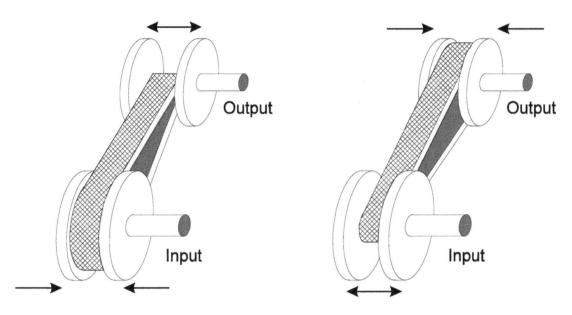

Figure 1.8:
Adjustable sheave belt-type mechanical VSD

The mechanical efficiency of this type of VSD is typically about 90% at maximum load. They are often used for machine tool or material handling applications. However, they are increasingly being superseded by small single phase AC or DC variable speed drives.

1.5.2 Metallic friction drives

Another type of mechanical drive is the metallic friction VSD unit, which can transmit power through the friction at the point of contact between two shaped or tapered wheels. Speed is adjusted by moving the line of contact relative to the rotation centers. Friction between the parts determines the transmission power and depends on the force at the contact point.

The most common type of friction VSD uses two rotating steel balls, where the speed is adjusted by tilting the axes of the balls. These can achieve quite high capacities of up to 100 kW and they have excellent speed repeatability. Speed ratios of 5:1 up to 25:1 are common.

To extend the life of the wearing parts, friction drives require a special lubricant that hardens under pressure. This reduces metal-to-metal contact, as the hardened lubricant is used to transmit the torque from one rotating part to the other.

1.6 Hydraulic variable speed drive methods

Hydraulic VSDs are often favored for conveyor drive applications because of the inherently *soft-start* capability of the hydraulic unit. They are also frequently used in all types of transportation and earthmoving equipment because of their inherently high starting torque. Both of the two common types work on the same basic principle where the prime mover, such as a fixed speed electric motor or a diesel/petrol engine, drives a hydraulic pump to transfer fluid to a hydraulic motor. The output speed can be adjusted by controlling the fluid flow rate or pressure. The two different types outlined below are characterized by the method employed to achieve the speed control.

1.6.1   Hydrodynamic types

Hydrodynamic variable speed couplings, often referred to as fluid couplings, are commonly used on conveyors. This type of coupling uses movable scoop tubes to adjust the amount of hydraulic fluid in the vortex between an impeller and a runner. Since the output is only connected to the input by the fluid, without direct mechanical connection, there is a slip of about 2% to 4%. Although this slip reduces efficiency, it provides good shock protection or soft-start characteristic to the driven equipment. The torque converters in the automatic transmissions of motor cars are hydrodynamic fluid couplings.

The output speed can be controlled by the amount of oil being removed by the scoop tube, which can be controlled by manual or automatic control systems. Operating speed ranges of up to 8:1 are common. A constant speed pump provides oil to the rotating elements.

1.6.2   Hydrostatic type

This type of hydraulic VSD is most commonly used in mobile equipment such as transportation, earthmoving and mining machinery. A hydraulic pump is driven by the prime mover, usually at a fixed speed, and transfers the hydraulic fluid to a hydraulic motor. The hydraulic pump and motor are usually housed in the same casing that allows closed circuit circulation of the hydraulic fluid from the pump to the motor and back.

The speed of the hydraulic motor is directly proportional to the rate of flow of the fluid and the displacement of the hydraulic motor. Consequently, variable speed control is based on the control of both fluid flow and adjustment of the pump and/or motor displacement. Practical drives of this type are capable of a very wide speed range, steplessly adjustable from zero to full synchronous speed.

The main advantages of hydrostatics VSDs, which make them ideal for earthmoving and mining equipment, are:

- High torque available at low speed
- High power-to-weight ratio
- The drive unit is not damaged even if it stalls at full load
- Hydrostatics VSDs are normally bi-directional

Output speed can be varied smoothly from about 40 rev/min to 1450 rev/min up to a power rating of about 25 kW. Speed adjustment can be done manually from a hand-wheel or remotely using a servo-motor. The main disadvantage is the poor speed holding capability. Speed may drop by up to 35 rev/min between 0% and 100% load.

Hydrostatic VSDs fall into four categories, depending on the types of pumps and motors.

- **Fixed displacement pump – fixed displacement motor**
  The displacement volume of both the pump and the motor is not adjustable. The output speed and power are controlled by adjusting a flow control valve located between the hydraulic pump and motor. This is the cheapest solution,

but efficiency is low, particularly at low speeds. So these are applied only where small speed variations are required.

- **Variable displacement pump – fixed displacement motor**
  The output speed is adjusted by controlling the pump displacement. Output torque is roughly constant relative to speed if pressure is constant. Thus power is proportional to speed. Typical applications include winches, hoists, printing machinery, machine tools and process machinery.

- **Fixed displacement pump – variable displacement motor**
  The output speed is adjusted by controlling the motor displacement. Output torque is inversely proportional to speed, giving a relatively constant power characteristic. This type of characteristic is suitable for machinery such as rewinders.

- **Variable displacement pump – variable displacement motor**
  The output speed is adjusted by controlling the displacement of the pump, motor or both. Output torque and power are both controllable across the entire speed range in both directions.

1.7 Electromagnetic or 'Eddy Current' coupling

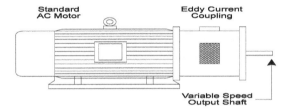

Figure 1.9:
Eddy current coupling mounted onto SCIM

The electromagnetic or 'Eddy Current' coupling is one of the oldest and simplest of the electrically controlled variable speed drives and has been used in industrial applications for over 50 years. In a similar arrangement to hydraulic couplings, eddy current couplings are usually mounted directly onto the flange of a standard squirrel cage induction motor between the motor and the driven load as shown in Figure 1.9.

Using the principles of electromagnetic induction, torque is transferred from a rotating drum, mounted onto the shaft of a fixed speed electric motor, across the air gap to an output drum and shaft, which is coupled to the driven load. The speed of the output shaft depends on the slip between the input and output drums, which is controlled by the magnetic field strength. The field winding is supplied with DC from a separate variable voltage source, which was traditionally a variac but is now usually a small single-phase thyristor converter.

There are several slightly different configurations using the electromagnetic induction principle, but the most common two constructions are shown in Figure 1.10. It comprises a cylindrical input drum and a cylindrical output drum with a small air gap between them. The output drum, which is connected to the output shaft, is capable of rotating freely relative to the input drum. A primary electromagnetic field is provided by a set of field coils that are connected to an external supply.

In configuration Figure 1.10(a), the field coils are mounted directly onto the rotating output drum, which then requires sliprings to transfer the excitation current to the field coils. On larger couplings, this arrangement can be difficult to implement and also sliprings create additional maintenance problems. In configuration Figure 1.10(b), the field coils are supported on the frame with the output drum closely surrounding it. This configuration avoids the use of sliprings.

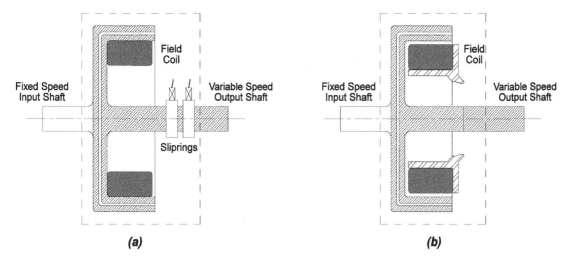

Figure 1.10:
Cross section of the eddy current couplings

(a) Field coils mounted onto the output drum

(b) Field coils mounted onto the fixed frame

The operating principle is based on the following:

- When a conducting material moves through the flux lines of a magnetic field, eddy currents are induced in the surface of the material, which flow in circular paths.
- The magnitude of the eddy currents is determined by the primary flux density and the rate at which the rotating part cuts these primary flux lines, i.e. the magnitude of the eddy currents depends on the magnetic field strength and the relative speed between the input and output shafts.
- These eddy currents collectively establish their own magnetic field which interacts with the primary magnetic flux in such a way as to resist the relative motion between them, thus providing a magnetic coupling between input and output drums.
- Consequently, torque can be transferred from a fixed speed prime mover to the output shaft, with some slip between them.
- The output torque and the slip are dependent on the strength of the electromagnetic field, which can be controlled from an external voltage source.

In the practical implementation, the input and output drums are made from a ferromagnetic material, such as iron, with a small air gap between them to minimize the leakage flux. The field coils, usually made of insulated copper windings, are mounted on the static part of the frame and are connected to a DC voltage source via a terminal box on the frame. Variable speed is obtained by controlling the field excitation current, by adjusting the voltage output of a small power electronic converter and control circuit. Speed adjustments can be made either manually from a potentiometer or remotely via a 4–20 mA control loop. An important feature of the eddy current coupling is the very low power rating of the field controller, which is typically 2% of the rated drive power.

When this type of drive is started by switching on the AC motor, the motor quickly accelerates to its full speed. With no voltage applied to the field coils, there are no lines of flux and no coupling, so the output shaft will initially be stationary. When an excitation current is applied to the field coils, the resulting lines of flux cut the rotating input drum at the maximum rate and produce the maximum eddy current effect for that field strength.

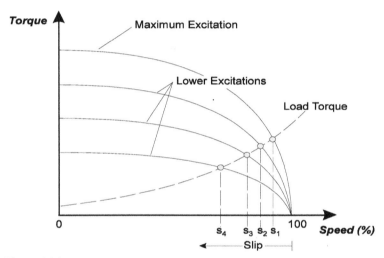

Figure 1.11:
Torque–speed curves for the eddy current coupling

The interaction between the primary flux and the secondary field produced by the eddy currents establishes an output torque, which accelerates the output shaft and the driven load. As the output drum accelerates, the relative speed between the two drums decreases and reduces the rate at which the lines of flux cut the rotating drum. The magnitude of the Eddy Currents and secondary magnetic field falls and, consequently, reduces the torque between them.

With a constant field excitation current, the output shaft will accelerate until the output torque comes into equilibrium with that of the driven machine. The output speed can be increased (reduce the slip) by increasing the field excitation current to increase the primary magnetic field strength. The output speed can be reduced (increase slip) by reducing the field excitation current.

To transfer torque through the interaction of two magnetic fields, eddy currents must exist to set up the secondary magnetic field. Consequently, there must always be a difference in speed, called the *slip*, between the input drum and the output drum. This behavior is very similar to that of the AC squirrel cage induction motor (SCIM) and indeed the same principles apply. The eddy current coupling produces a torque–speed curve quite similar to a SCIM as shown in Figure 1.11.

Theoretically, the eddy current coupling should be able to provide a full range of output speeds and torques from zero up to just below the rated speed and torque of the motor, allowing of course for slip. In practice, this is limited by the amount of torque that can be transferred continuously through the coupling without generating excessive heat.

When stability is reached between the motor and the driven load connected by an *Eddy Current coupling*, the output torque on the shaft is equal to the input torque from the AC motor. However, the speeds of the input and output shafts will be different due to the slip. Since power is a product of torque and speed, the difference between the input and output power, the losses, appears as heat in the coupling. These losses are dissipated through cooling fins on the rotating drums.

These losses may be calculated as follows:

$$Losses = (P_1 - P_0) \quad \text{kW}$$

$$Losses = \frac{T(n_1 - n_2)}{9550} \quad \text{kW}$$

The worst case occurs at starting, with the full rated torque of the motor applied to the driven load at zero output speed, the losses in the coupling are the full rated power of the motor. Because of the difficulty of dissipating this amount of energy, in practice it is necessary to limit the continuous torque at low speeds.

Alternatively, some additional cooling may be necessary for the coupling, but this results in additional capital costs and low energy efficiency. In these cases, other types of VSDs may be more suitable. Consequently, an eddy current coupling is most suited to those types of driven load, which have a low torque at low speed, such as centrifugal pumps and fans. The practical loadability of the eddy current coupling is shown in the figure below.

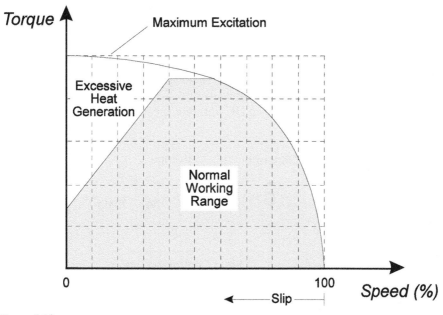

Figure 1.12:
Loadability of the eddy current coupling

A major drawback of the eddy current coupling is its poor dynamic response. Its ability to respond to step changes in the load or the speed setpoint depends on the time constants associated with the highly inductive field coil, the eddy currents in the ferro-magnetic drums and the type of control system used. The field coil time constant is the most significant factor and there is very little that can be done to improve it, except possibly to use a larger coupling. Closed loop speed control with tachometer feedback can also be used to improve its performance. But there are many applications where the dynamic response or output speed accuracy are not important issues and the eddy current coupling has been proven to be a cost effective and reliable solution for these applications.

1.8   Electrical variable speed drive methods

In contrast to the mechanical and hydraulic variable speed control methods, electrical variable speed drives are those in which the speed of the electric motor itself, rather than an intermediary device, is controlled. Variable speed drives that control the speed of DC motors are loosely called *DC variable speed drives* or simply *DC drives* and those that control the speed of AC motors are called *AC variable speed drives* or simply *AC drives*. Almost all electrical VSDs are designed for operation from the standard 3-phase AC power supply system.

Historically, two of the best known electrical VSDs were the schrage motor and the Ward-Leonard system. Although these were both designed for operation from a 3-phase AC power supply system, the former is an AC commutator motor while the latter uses a DC generator and motor to effect speed control.

1.8.1   AC commutator motor – schrage motor

The schrage motor is an AC commutator motor having its primary winding on the rotor. The speed was changed by controlling the position of the movable brushes by means of a hand-wheel or a servo-motor. Although it was very popular in its time, this type of motor is now too expensive to manufacture and maintain and is now seldom used.

1.8.2   Ward-Leonard system

The Ward-Leonard system comprises a fixed speed 3-phase AC induction motor driving a separately excited DC generator that, in turn, feeds a variable voltage to a shunt wound DC motor. So this is essentially a DC variable speed drive.

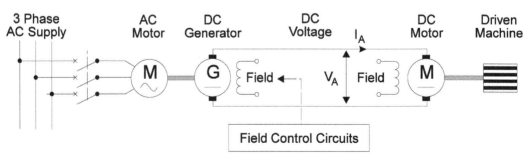

Figure 1.13:
The Ward-Leonard system

DC drives have been used for variable speed applications for many decades and historically were the first choice for speed control applications requiring accurate speed control, controllable torque, reliability and simplicity. The basic principle of a DC variable speed drive is that the speed of a separately excited DC motor is directly proportional to the voltage applied to the armature of the DC motor. The main changes over the years have been concerned with the different methods of generating the variable DC voltage from the 3-phase AC supply.

In the case of the Ward-Leonard system, the output voltage of the DC generator, which is adjusted by controlling the field voltage, is used to control the speed of the DC motor as shown in Figure 1.13. This type of variable speed drive had good speed and torque characteristics and could achieve a speed range of 25:1. It was commonly used for winder drives where torque control was important. It is no longer commonly used because of the high cost of the 3 separate rotating machines. In addition, the system requires considerable maintenance to keep the brushes and commutators of the two DC machines in good condition.

In modern DC drives, the motor-generator set has been replaced by a thyristor converter. The output DC voltage is controlled by adjusting the **firing angle** of the thyristors connected in a bridge configuration connected directly to the AC power supply.

1.8.3   Electrical variable speed drives for DC motors (DC drives)

Since the 1970s, the controlled DC voltage required for DC motor speed control has been more easily produced from the 3-phase AC supply using a *static* power electronic AC/DC converter, or sometimes called a **controlled rectifier**. Because of its low cost and low maintenance, this type of system has completely superseded the Ward-Leonard system. There are several different configurations of the AC/DC converter, which may contain a full-wave 12-pulse bridge, a full-wave 6-pulse bridge or a half-wave 3-pulse bridge. On larger DC drive systems, 12-pulse bridges are often used.

The most common type of AC/DC converter, which meets the steady state and dynamic performance requirements of most VSD applications, comprises a 6-pulse thyristor bridge, electronic control circuit and a DC motor as shown Figure 1.14. The 6-pulse bridge produces less distortion on the DC side than the 3-pulse bridge and also results in lower losses in the DC motor. On larger DC drive systems, 12-pulse bridges are often used to reduce the harmonics in the AC power supply system.

The efficiency of an AC/DC converter is high, usually in excess of 98%. The overall efficiency of the DC drive, including the motor, is lower and is typically about 90% at full load depending on the size of the motor. The design and performance of power electronic converters is described in detail in Chapter 3: Power electronic converters.

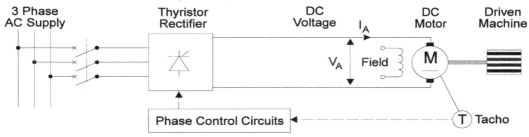

Figure 1.14:
Basic construction of a 6-pulse DC variable speed drive

AC/DC converters of this type are relatively simple and robust and can be built for VSDs of up to several megaWatts with good control and performance characteristics. Since the DC motor is relatively complex and expensive, the main disadvantage of this type of VSD in comparison to an AC VSD, is the reliability of the DC motor. Although the maintenance requirements of a DC motor are inherently higher than an AC induction motor, provided that the correct brush grade is used for the speed and current rating, the life of the commutator and brushgear can be quite long and maintenance minimal.

The fundamental principles of a DC variable speed drive, with a shunt wound DC motor, are relatively easy to understand and are covered by a few simple equations as follows:

- The armature voltage V_A is the sum of the internal armature EMF V_E and the volt drop due to the armature current I_A flow through the armature resistance R_A.

  Armature Voltage $\quad V_A = V_E + I_A R_A$

- The DC motor speed is directly proportional to the armature back EMF V_E and indirectly proportional to the field flux Φ, which in turn depends on the field excitation current I_E. Thus, the rotational speed of the motor can be controlled by adjusting either the armature voltage, which controls V_E, or the field current, which controls the Φ.

  Motor Speed $\quad\quad n \propto \dfrac{V_E}{\Phi}$

- The output torque T of the motor is proportional to the product of the armature current and the field flux.

  Output Torque $\quad T \propto I_A \Phi$

- The direction of the torque and direction of rotation of the DC motor can be reversed either by changing the polarity of Φ, called field reversal, or by changing the polarity of I_A, called armature current reversal. These can be achieved by reversing the supply voltage connections to the field or to the armature.

- The output power of the motor is proportional to the product of torque and speed.

  Output Power $\quad P \propto T n$

From these equations, the following can be deduced about a DC motor drive:

- The speed of a DC motor can be controlled by adjusting either the armature voltage or the field flux or both. Usually the field flux is kept constant, so the motor speed is increased by increasing the armature voltage.
- When the armature voltage V_A has reached the maximum output of the converter, additional increases in speed can be achieved by reducing the field flux. This is known as the *field weakening* range. In the field weakening range, the speed range is usually limited to about 3:1, mainly to ensure stability and continued good commutation.
- The motor is able to develop its full torque over the normal speed range. Since torque is not dependent on V_A, the full-load torque output is possible over the normal speed range, even at standstill (zero speed).
- The output power is zero at zero speed. In the normal speed range and at constant torque, the output power increases in proportion to the speed.

• In the field weakening range, the motor torque falls in proportion to the speed. Consequently, the output power of the DC motor remains constant.

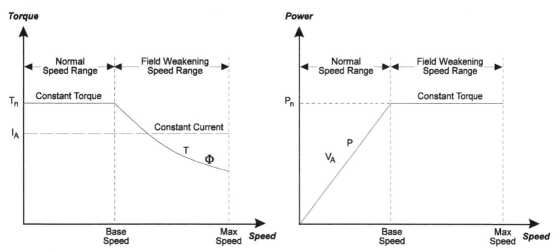

Figure 1.15:
Torque and power of a DC drive over the speed range

Although a DC machine is well suited for adjustable speed drive applications, there are some limitations due to the mechanical commutator and brushes, which:

• Impose restrictions on the ambient conditions, such as temperature and humidity
• Are subject to wear and require periodic maintenance
• Limit the maximum power and speed of machines that can be built

1.8.4   Electrical variable speed drives for AC motors (AC drives)

One of the lingering problems with thyristor controlled DC drives is the high maintenance requirement of the DC motor. Since the 1980s, the popularity of AC variable speed drives has grown rapidly, mainly due to advances in power electronics and digital control technology affecting both the cost and performance of this type of VSD. The main attraction of the AC VSDs is the rugged reliability and low cost of the **squirrel cage AC induction motor** compared to the DC motor.

In the AC VSD, the mechanical commutation system of the DC motor, has been replaced by a power electronic circuit called the *inverter*. However, the main difficulty with the AC variable speed drive has always been the complexity, cost and reliability of the AC frequency inverter circuit.

The development path from the Ward-Leonard system to the thyristor controlled DC drive and then to the PWM-type AC variable voltage variable frequency converter is illustrated in Figure 1.16. In the first step from (a) to (b), the high cost motor-generator set has been replaced with a phase-controlled thyristor rectifier.

In the second step from (b) to (d), the high cost DC motor has been replaced with a power electronic PWM inverter and a simple rugged AC induction motor. Also, the rectifier is usually a simple diode rectifier.

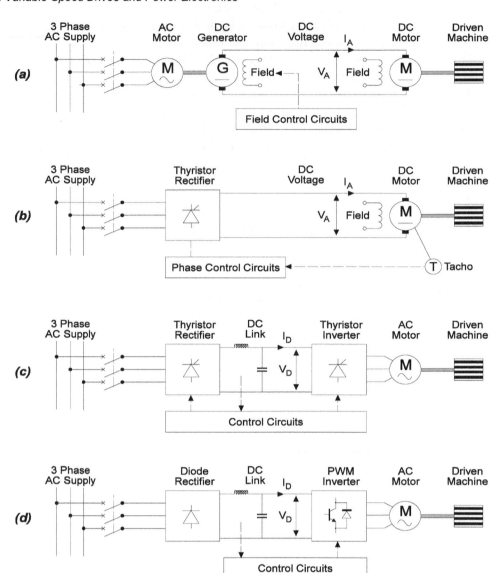

Figure 1.16:
Main components of various types of variable speed drive:

    (a) Ward-Leonard system

    (b) Thyristor controlled DC drive

    (c) Voltage source inverter (PAM) AC drive

    (d) PWM voltage source (PWM) AC drive

Frequency control, as a method of changing the speed of AC motors, has been a well known technique for decades, but it has only recently become a technically viable and economical method of variable speed drive control. In the past, DC motors were used in most variable speed drive applications in spite of the complexity, high cost and high maintenance requirements of the DC motors. Even today, DC drives are still often used for the more demanding variable speed drive applications. Examples of this are the sectional drives for paper machines, which require fast dynamic response and separate control of speed and torque.

Developments in power electronics over the last 10 to 15 years has made it possible to control not only the speed of AC induction motors but also the torque. Modern AC variable speed drives, with *flux-vector control,* can now meet all the performance requirements of even the most demanding applications.

In comparison to DC drives, AC drives have become a more cost effective method of speed control for most variable speed drive applications up to 1000 kW. It is also the technically preferred solution for many industrial environments where reliability and low maintenance associated with the AC squirrel cage induction motor are important.

The fundamental principles of an AC variable speed drive are relatively easy to understand and are covered by a few simple equations as follows:

- The speed (*n*) of the motor can be controlled either by adjusting the supply frequency (*f*) or the number of poles (*p*). In an AC induction motor, the synchronous speed, which is the speed at which the stator field rotates, is governed by the simple formula:

Synchronous Speed $\qquad n_\text{S} = \dfrac{120f}{p} \text{ rev/min}$

Although there are special designs of induction motors, whose speed can be changed in one or more steps by changing the number of poles, it is impractical to continuously vary the number of poles to effect smooth speed control. Consequently, the fundamental principle of modern AC variable speed drives is that the speed of a fixed pole AC induction motor is proportional to the *frequency* of the AC voltage connected to it.

In practice, the actual speed of the rotor shaft is slower than the synchronous speed of the rotating stator field, due to the **slip** between the stator field and the rotor. This is covered in detail in Chapter 2: 3-Phase AC induction motors.

Actual speed $\quad n = (n_\text{s} - slip) \text{ rev/min}$

The slip between the synchronous rotating field and the rotor depends on a number of factors, being the stator voltage, the rotor current and the mechanical load on the shaft. Consequently, the speed of an AC induction motor can also be adjusted by controlling the slip of the rotor relative to the stator field. Slip control is discussed in Section 1.8.5.

Unlike a shunt wound DC motor, the stator field flux in an induction motor is also derived from the supply voltage and the flux density in the air gap will be affected by changes in the frequency of the supply voltage. The air-gap flux (Φ) of an AC induction motor is directly proportional to the magnitude of the supply voltage (*V*) and inversely proportional to the frequency (*f*).

Air-gap Flux $\quad \Phi \propto \dfrac{V}{f}$

To maintain a constant field flux density in the metal parts during speed control, the stator voltage must be adjusted in proportion to the frequency. If not and the flux density is allowed to rise too high, saturation of the iron parts of the motor will result in high excitation currents, which will cause excessive losses and heating. If the flux density is allowed to fall too low, the output torque will drop and affect the performance of the AC

Drive. Air-gap flux density is dependent on both the frequency and the magnitude of the supply voltage.

So the speed control of AC motors is complicated by the fact that both voltage and frequency need to be controlled simultaneously, hence the name variable voltage, variable frequency (VVVF) converter.

- In a similar way to the DC motor, the output torque of the AC motor depends on the product of the air-gap flux density and the rotor current I_R. So, to maintain constant motor output torque, the flux density must be kept constant which means that the ratio V/f must be kept constant.

$$\text{Output Torque} \qquad T \propto \Phi\, I_R \quad \text{Nm}$$

- The direction of rotation of the AC motor can be reversed by changing the firing sequence power electronic valves of the inverter stage. This is simply done through the electronic control circuit.
- Output power of the AC motor is proportional to the product of torque and speed.

$$\text{Output Power} \qquad P \propto T\, n \quad \text{kW}$$

The basic construction of a modern AC frequency converter is shown in the figure below.

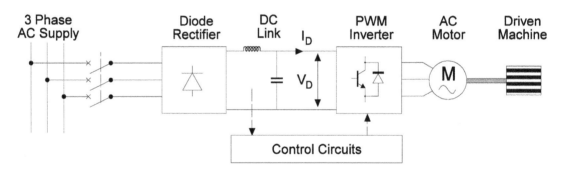

Figure 1.17:
Main components of a typical PWM-type AC drive

The mains AC supply voltage is converted into a DC voltage and current through a rectifier. The DC voltage and current are filtered to smooth out the peaks before being fed into an inverter, where they are converted into a variable AC voltage and frequency. The output voltage is controlled so that the ratio between voltage and frequency remains constant to avoid over-fluxing the motor. The AC motor is able to provide its rated torque over the speed range up to 50 Hz without a significant increase in losses.

The motor can be run at speeds above rated frequency, but with reduced output torque. Torque is reduced as a result of the reduction in the air-gap flux, which depends on the *V/f* ratio. The locus of the induction motor torque–speed curves are at various frequencies are shown in the figure below. At frequencies below 50 Hz, a *constant torque* output from the motor is possible. At frequencies above the base frequency of 50 Hz, torque is reduced in proportion to the reduction in speed.

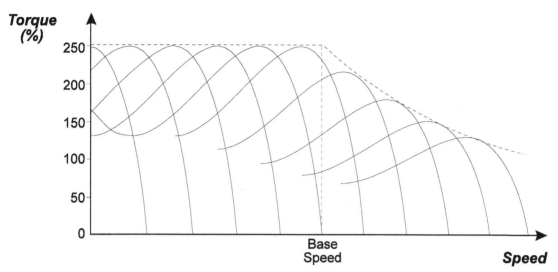

Figure 1.18:
Locus of the motor torque-speed curves at various frequencies

One of the main advantages of this VVVF speed control system is that, whilst the controls are necessarily complex, the motors themselves can be of squirrel cage construction, which is probably the most robust, and maintenance free form of electric motor yet devised. This is particularly useful where the motors are mounted in hazardous locations or in an inaccessible position, making routine cleaning and maintenance difficult. Where a machine needs to be built into a flameproof, or even waterproof enclosure, this can be done more cheaply with a squirrel cage AC induction motor than for a DC motor.

On the other hand, an additional problem with standard AC squirrel cage motors, when used for variable speed applications, is that they are cooled by means of a shaft mounted fan. At low speeds, cooling is reduced, which affects the *loadability* of the drive. The continuous output torque of the drive must be derated for lower speeds, unless a separately powered auxiliary fan is used to cool the motor. This is similar to the cooling requirements of DC motors, which require a separately powered auxiliary cooling fan.

From the equations above, the following deductions can be made about an AC drive:

- The speed of an AC induction motor can be controlled by adjusting the frequency and magnitude of the stator voltage. Motor speed is proportional to frequency, but the voltage must be simultaneously adjusted to avoid over-fluxing the motor.
- The AC motor is able to develop its full torque over the normal speed range, provided that the flux is held constant, (V/f ratio kept constant). A standard AC motor reaches its rated speed, when the frequency has been increased to rated frequency (50 Hz) and stator voltage V has reached its rated magnitude.
- The speed of an AC induction motor can be increased above its nominal 50 Hz rating, but the V/f ratio will fall because the stator voltage cannot be increased any further. This results in a fall of the air-gap flux and a reduction in output torque. As with the DC motor, this is known as the *field weakening* range. The performance of the AC motor in the field weakening range is similar to that of the DC motor and is characterized by constant power, reduced torque.

- The output power is zero at zero speed. In the normal speed range and at constant torque, the output power increases in proportion to the speed.
- In the field weakening range, the motor torque falls in proportion to the speed and the output power of the AC motor remains constant.

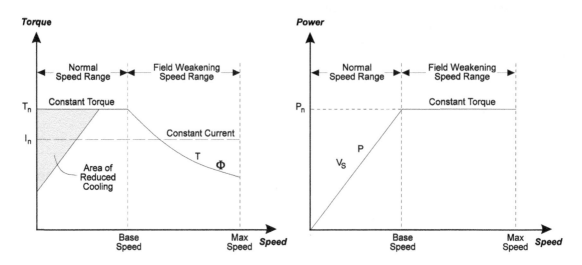

Figure 1.19:
Torque and power of an AC drive over the speed range

1.8.5   Slip control AC variable speed drives

When an AC induction motor is started direct-on-line (DOL), the electrical power supply system experiences a current surge which can be anywhere between 4 to 10 times the rated current of the motor. The level of inrush current depends on the design of the motor and is independent of the mechanical load connected to the motor. A standard squirrel cage induction motor has an inrush current typically of 6 times the rated current of the motor. The starting torque, associated with the inrush current, is typically between 1.5 to 2.5 times the rated torque of the motor. When the rotor is stationary, the slip is 100% and the speed is zero. As the motor accelerates, the slip decreases and the speed eventually stabilizes at the point where the motor output torque equals the mechanical load torque, as illustrated in Figures 1.20 and 1.22.

The basic design of a squirrel cage induction motor (SCIM) and a wound rotor induction motor (WRIM) are very similar, the main difference being the design and construction of the rotor. The design and performance of AC induction motors is described in considerable detail in Chapter 2: 3-Phase AC induction motors. In AC induction motors, the slip between the synchronous rotating stator field and the rotor is mainly dependent on the following two factors, either of which can be used to control the motor speed:

- **Stator voltage:** Affects both the flux and the rotor current.
- **Rotor current:** For a SCIM, this depends on the rotor design.

       For a WRIM, this depends on the external rotor connections

Stator voltage control

The reduction of the AC supply voltage to an induction motor has the effect of reducing both the air-gap flux (Φ) and the rotor current (I_R). The output torque of the motor behaves in accordance with the following formula:

$$\text{Output Torque} \quad T \propto \Phi I_R \quad \text{Nm}$$

Since both Φ and I_R decrease with the voltage, the output torque of the motor falls roughly as the square of the voltage reduction. So when voltage is reduced, torque decreases, slip increases and speed decreases. The characteristic curves in Figure 1.20 show the relationship between torque and speed for various values of the supply voltage.

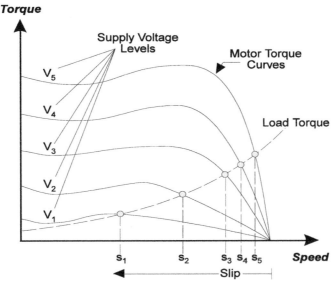

Figure 1.20:
Torque–speed curves of an induction motor with reduced supply voltage

V_1 = *Low level of supply voltage*

V_4 = *High level of supply voltage*

V_5 = *Full rated supply voltage*

From this figure, the speed stabilizes at the point where the motor torque curve, for that voltage, intersects with the load–torque curve. The application of this technique for speed control is very limited because the resulting speed is dependent on the mechanical load torque. Consequently, speed holding is poor unless speed feedback is used, for example by installing a shaft encoder or tachometer on the motor.

Reduced voltage control is not usually for speed control in industry, but for motor torque control, mainly for *soft starting* squirrel cage induction motors. Reduced voltage (reduced torque) soft starting has the following main advantages:

- Reduces mechanical shock on the driven machinery, hence the name soft starting
- Reduces the starting current surge in the electrical power supply system
- Reduces water hammer during starting and stopping in pumping systems

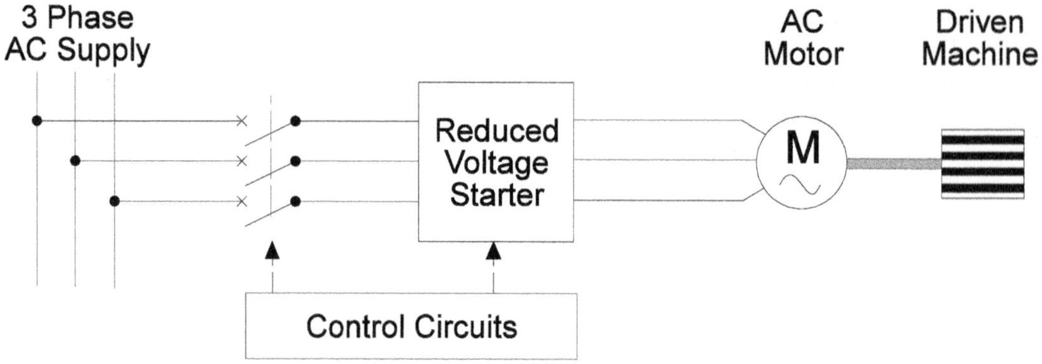

Figure 1.21:
Typical connections of a reduced voltage starter with an SCIM

The following devices are commonly used in industry for **reduced voltage starting** and are typically connected as shown in Figure 1.21:

- Auto-transformers in series with stator: Reactive volt drop
- Reactors in series with stator: Reactive volt drop
- Resistors in series with stator: Resistive volt drop.
- Thyristor bridge with electronic control: Chopped voltage waveform

Referring to the above, the characteristics of stator voltage control are as follows:

- Starting current inrush decreases as the square of the reduction in supply voltage.
- Motor output torque decreases as the square of the reduction in supply voltage.
- For reduced stator voltage, starting torque is always lower than DOL starting torque.
- Reduced voltage starting is not suitable for applications that require a high break-away torque.
- Stator voltage control is not really suitable for speed control because of poor speed holding capability.

Rotor current control

Rotor current control is another effective method of slip control that has successfully been used with induction motors for many decades. With full supply voltage on the stator, giving a constant flux Φ, the rotor current I_R can be controlled by adjusting the effective rotor resistance R_R.

For this method of control, it is necessary to have access to the 3-phase rotor windings. A special type of induction motor, known as a *wound rotor induction motor (WRIM)*, sometimes also called a *slipring motor*, is used for these applications. In a WRIM, the connections to the rotor windings are brought out to terminals via 3 slip-rings and brushes, usually mounted at the non-drive end of the shaft. By connecting external resistance banks to the rotor windings, the rotor current can be controlled. Since output torque is proportional to the product of Φ and I_R, with a constant field flux, the rotor current affects the torque–speed characteristic of the motor as shown in Figure 1.22.

Increasing the rotor resistance reduces the rotor current and consequently the output torque. With lower output torque, the slip increases and speed decreases.

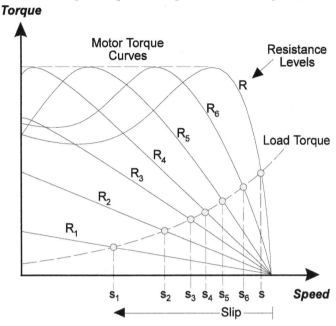

Figure 1.22:
Torque–speed curves of a WRIM with external rotor resistance

R_1 = *High external rotor resistance connected*

R_6 = *Low external rotor resistance connected*

R = *Normal rotor resistance*

As with stator voltage control, this method of speed control has a number of limitations. The speed holding capability, for changes in mechanical load, is poor. Again, this method of control is more often used to control starting torque rather than for speed control. In contrast to stator voltage control, which has low starting torque, *rotor current control* can provide a high starting torque with the added advantage of soft starting.

The following devices are commonly used in industry for rotor current control:
- Air-cooled resistor banks with bypass contactors
- Oil-cooled resistor banks with bypass contactors
- Liquid resistor starters with controlled depth electrodes
- Thyristor converters for rotor current control and slip energy recovery

Some of the characteristics of rotor current control are as follows:
- Starting current inrush is reduced in direct proportion to the rotor resistance.
- Starting torque, for certain values of rotor resistance, is higher than DOL starting torque and can be as high as the breakdown torque.
- Starting with a high external rotor resistance, as resistance is decreased in steps, the starting torque is progressively increased from a low value up to the breakdown torque.

- This type of starting is ideal for applications that require a high pull-away torque with a soft start, such as conveyors, crushers, ball mills, etc.
- Rotor current control is not ideal for speed control because of poor speed holding capability. However, it can be used for limited speed control, provided the speed range is small, typically 70% to 100% of motor rated speed. Motor speed holding is improved with the use of a shaft encoder or tachometer.

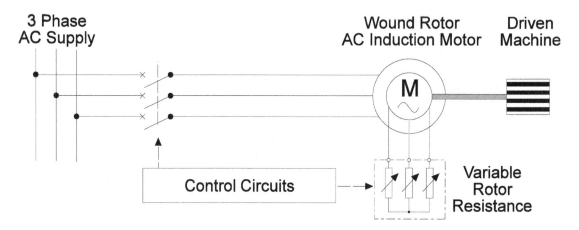

Figure 1.23:
Typical connections of a WRIM with rotor resistance starter

When the rotor current is controlled by external rotor resistors, a considerable amount of heat, known as the slip energy, needs to be dissipated in the resistor banks. In practice, rotor resistors are used for starting large induction motors and to accelerate heavy mechanical loads up to full speed. At full speed, the resistors are bypassed by means of contactors and the motor runs with a shorted rotor. Consequently, these losses occur for relatively short periods of time and are not considered to be of major significance.

However, when rotor resistors are used for speed control over an extended period of time, the energy losses can be high and the overall efficiency of the drive low. At constant output torque, the energy losses in the resistors are directly proportional to the slip. So as the speed is decreased, the efficiency decreases in direct proportion. For example, a WRIM running at 70% of rated speed at full load will need to dissipate roughly 30% of its rated power in the rotor resistors.

Slip energy recovery system

The slip energy recovery (SER) system is a further development of rotor current control, which uses power electronic devices, instead of resistors, for controlling the rotor current. The main components of the slip energy recovery system are shown in Figure 1.24. The rotor current is controlled by adjusting the firing angle of the rectifier bridge. With the rectifier bridge turned off, the rotor current is zero and with the thyristor bridge full on, the rotor current approaches rated current. The rectifier bridge can be controlled to provide any current between these outer limits. Instead of dumping the slip energy into a resistor, it is smoothed through a large choke and converted back into 3-phase AC currents, which are pumped back into the mains at 50 Hz through a matching transformer. The thyristors of the rectifier bridge are commutated by the rotor voltage, while the thyristors of the inverter bridge are commutated by the supply voltage. The DC link allows the two sides of the converter to run at different frequencies. The tacho is used for speed feedback to improve the speed holding capability of this variable speed drive.

Using SER technique, the slip energy losses can be recovered and returned to the power supply system, thus improving the efficiency of the drive.

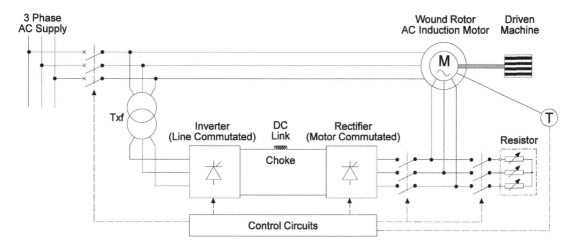

Figure 1.24:
The main components of a slip energy recovery system

Some interesting aspects of the slip energy recovery system are as follows:

- The rotor connected SER converter need only be rated for the slip energy, which depends on the required speed range. For example, for a speed range of 80% to 100% of rated motor speed, the SER converter should be rated at roughly 20% of motor power rating. If the speed range needs to be broadened to 70% to 100%, the rating of the SER converter needs to be increased to roughly 30% of motor power rating. In contrast, stator connected VVVF converters, commonly used for the speed control of squirrel cage induction motors, need to be rated for >100% of the motor power rating.

- Because the SER converter rating is lower than motor rating, the slip power at starting would exceed the rating of the converter. It has become common practice to use an additional rotor resistance starter, selected by contactors from the control circuit, for the starting period from standstill. These resistors can be air-cooled, oil-cooled or the liquid type. Once the WRIM motor has been accelerated up to the variable speed range, the SER converter is connected and the resistors disconnected. These resistors have the added advantage of providing a standby solution in the event of a SER converter failure, when the motor can be started and run at fixed speed without the SER system.

- For additional flexibility, a bypass contactor is usually provided to short circuit the rotor windings and allow the motor to run at fixed speed.

The slip energy recovery system is most often used by large water supply authorities for soft starting and limited speed control of large centrifugal pumps, typically 1 MW to 10 MW. In these applications, they are a more cost effective solution than the equivalent stator connected AC or DC drives. Another increasingly common application is the starting and limited speed control (70% to 100%) of large SAG mills in mineral processing plants, typically 1 MW to 5 MW.

1.8.6   Cycloconverters

A cycloconverter is a converter that synthesizes a 3-phase AC variable frequency output directly from a fixed frequency 3-phase AC supply, without going via a DC link. The cycloconverter is not new and the idea was developed over 50 years ago using mercury arc rectifiers.

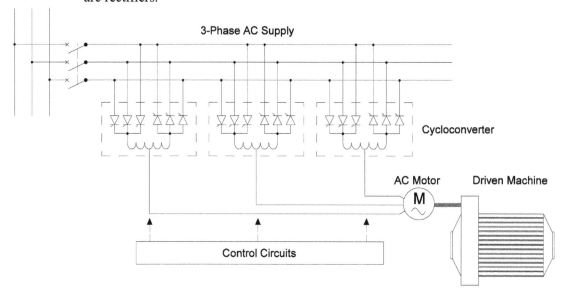

Figure 1.25:
The main components of a cycloconverter

The low frequency AC waveform is produced using two back-to-back thyristors per phase, which are allowed to conduct alternatively. By suitable phase angle control, the output voltage and load current can be made to change in magnitude and polarity in cyclic fashion. The main limitation of the cycloconverter is that it cannot generate frequencies higher than the AC supply frequency. In fact, a frequency of about 30% of the supply frequency is the highest practically possible with reasonable waveforms. The lower the frequency, the better the waveform. The system is inherently capable of regeneration back into the mains.

The cycloconverter requires a large number of thyristors, and the control circuitry is relatively complex but, with the advent of microprocessors and digital electronics, the implementation of the control circuits has become more manageable.

Because of the low frequency output, cycloconverters are suited mainly for large slow speed drives, where it is used to drive either a large induction motor or a synchronous motor. Typical applications are SAG or ball mills, rotary cement kilns, large crushers, mine-winders, etc.

1.8.7   Servo-drives

Servo-drives are used in those drive applications which require a high level of precision, usually at relatively low powers. This often includes rapid stop-start cycles, very high acceleration torques, accurate positioning with controllable velocity and torque profiles.

The use of servo-drives for industrial manufacturing and materials handling has also become far more common, particularly for accurate positioning systems. This type of drive differs from a normal open loop VVVF drive in the following respects:

- Accuracy and precision of the motor speed and torque output are far in excess of what is normally possible with AC induction motors
- A servo-motor is usually designed to operate with a specific servo-converter
- Response of the servo-drive system to speed change demand is extremely fast
- Servo-drives provide full torque holding at zero speed
- Servo-drive inertia is usually very low to provide rapid response rates

Servo-drives are beyond the scope of this book and will not be covered here.

2

3-Phase AC induction motors

2.1 Introduction

For industrial and mining applications, 3-phase AC induction motors are the prime movers for the vast majority of machines. These motors can be operated either directly from the mains or from adjustable frequency drives. In modern industrialized countries, more than half the total electrical energy used in those countries is converted to mechanical energy through *AC induction motors*. The applications for these motors cover almost every stage of manufacturing and processing. Applications also extend to commercial buildings and the domestic environment. They are used to drive pumps, fans, compressors, mixers, agitators, mills, conveyors, crushers, machine tools, cranes, etc, etc. It is not surprising to find that this type of electric motor is so popular, when one considers its simplicity, reliability and low cost.

In the last decade, it has become increasingly common practice to use 3-phase squirrel cage AC induction motors with *variable voltage variable frequency (VVVF) converters* for variable speed drive (VSD) applications. To clearly understand how the VSD system works, it is necessary to understand the principles of operation of this type of motor.

Although the basic design of induction motors has not changed very much in the last 50 years, modern insulation materials, computer based design optimization techniques and automated manufacturing methods have resulted in motors of smaller physical size and lower cost per kW. International standardization of physical dimensions and frame sizes means that motors from most manufacturers are physically interchangeable and they have similar performance characteristics.

The reliability of squirrel cage AC induction motors, compared to DC motors, is high. The only parts of the squirrel cage motor that can wear are the bearings. Sliprings and brushes are not required for this type of construction. Improvements in modern pre-lubricated bearing design have extended the life of these motors.

Although **single-phase** AC induction motors are quite popular and common for low power applications up to approx 2.2 kW, these are seldom used in industrial and mining applications. Single-phase motors are more often used for domestic applications.

The information in this chapter applies mainly to 3-phase squirrel cage AC induction motors, which is the type most commonly used with VVVF converters.

2.2 Basic construction

The AC induction motor comprises 2 electromagnetic parts:

- Stationary part called the *stator*
- Rotating part called the *rotor*, supported at each end on bearings

The stator and the rotor are each made up of:

- An *electric circuit*, usually made of insulated copper or aluminum, to carry current
- A *magnetic circuit*, usually made from laminated steel, to carry magnetic flux

2.2.1 The stator

The *stator* is the outer stationary part of the motor, which consists of:

- The **outer cylindrical frame** of the motor, which is made either of welded sheet steel, cast iron or cast aluminum alloy. This may include feet or a flange for mounting.
- The **magnetic path**, which comprises a set of slotted steel laminations pressed into the cylindrical space inside the outer frame. The magnetic path is laminated to reduce eddy currents, lower losses and lower heating.
- A set of **insulated electrical windings**, which are placed inside the slots of the laminated magnetic path. The cross-sectional area of these windings must be large enough for the power rating of the motor. For a 3-phase motor, 3 sets of windings are required, one for each phase.

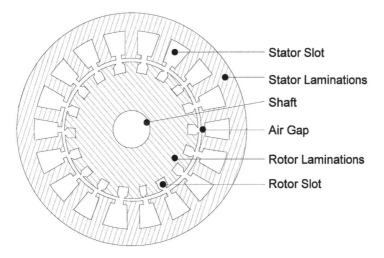

Figure 2.1:
Stator and rotor laminations

2.2.2   The rotor

This is the rotating part of the motor. As with the stator above, the rotor consists of a set of slotted steel laminations pressed together in the form of a cylindrical magnetic path and the electrical circuit. The electrical circuit of the rotor can be either:

- **Wound rotor type**, which comprises 3 sets of insulated windings with connections brought out to 3 sliprings mounted on the shaft. The external connections to the rotating part are made via brushes onto the sliprings. Consequently, this type of motor is often referred to as a *slipring motor*.
- **Squirrel cage rotor type**, which comprises a set of copper or aluminum bars installed into the slots, which are connected to an end-ring at each end of the rotor. The construction of these rotor windings resembles a '*squirrel cage*'. Aluminum rotor bars are usually die-cast into the rotor slots, which results in a very rugged construction. Even though the aluminum rotor bars are in direct contact with the steel laminations, practically all the rotor current flows through the aluminum bars and not in the laminations.

2.2.3   The other parts

The other parts, which are required to complete the induction motor are:

- *Two end-flanges* to support the two bearings, one at the drive-end (DE) and the other at the non drive-end (NDE)
- *Two bearings* to support the rotating shaft, at DE and NDE
- *Steel shaft* for transmitting the torque to the load
- *Cooling fan* located at the NDE to provide forced cooling for the stator and rotor
- *Terminal box* on top or either side to receive the external electrical connections

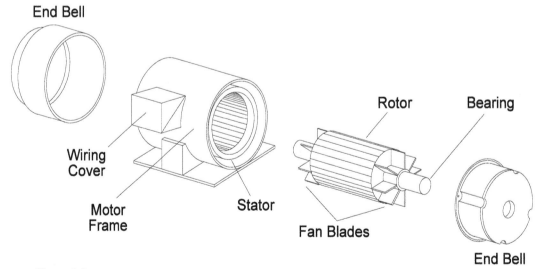

Figure 2.2:
Assembly details of a typical AC induction motor

2.3   Principles of operation

When a 3-phase AC power supply is connected to the stator terminals of an induction motor, 3-phase alternating currents flow in the stator windings. These currents set up a changing magnetic field (flux pattern), which rotates around the inside of the stator. The speed of rotation is in synchronism with the electric power frequency and is called the *synchronous speed*.

In the simplest type of 3-phase induction motor, the rotating field is produced by 3 fixed stator windings, spaced 120° apart around the perimeter of the stator. When the three stator windings are connected to the 3-phases power supply, the flux completes one rotation for every *cycle* of the supply voltage. On a 50 Hz power supply, the stator flux rotates at a speed of 50 revolutions per second, or 50 × 60 = 3000 rev per minute.

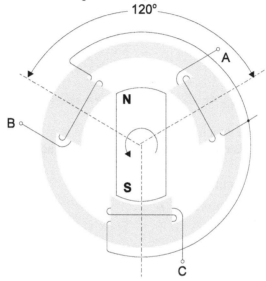

Figure 2.3:
Basic (simplified) principle of a 2 pole motor

A motor with only one set of stator electrical windings per phase, as described above, is called a *2 pole motor (2p)* because the rotating magnetic field comprises 2 rotating poles, one North-pole and one South-pole. In some countries, motors with 2 rotating poles are also sometimes called a *1 pole-pair motor*.

If there was a permanent magnet inside the rotor, it would follow in synchronism with the rotating magnetic field. The rotor magnetic field interacts with the rotating stator flux to produce a rotational force. A permanent magnet is only being mentioned because the principle of operation is easy to understand. The magnetic field in a normal induction motor is induced across the rotor air-gap as described below.

If the three windings of the stator were re-arranged to fit into half of the stator slots, there would be space for another 3 windings in the other half of the stator. The resulting rotating magnetic field would then have 4 poles (two North and two South), called a *4 pole motor*. Since the rotating field only passes 3 stator windings for each power supply cycle, it will rotate at half the speed of the above example, 1500 rev/min.

Consequently, induction motors can be designed and manufactured with the number of stator windings to suit the base speed required for different applications:

- 2 pole motors, stator flux rotates at 3000 rev/min
- 4 pole motors, stator flux rotates at 1500 rev/min
- 6 pole motors, stator flux rotates at 1000 rev/min

- 8 pole motors, stator flux rotates at 750 rev/min
- etc

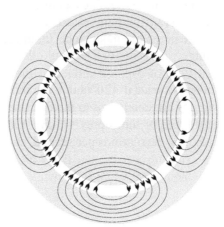

Figure 2.4:
Flux distribution in a 4 pole machine at any one moment

The speed at which the stator flux rotates is called the *synchronous speed* and, as shown above, depends on the number of poles of the motor and the power supply frequency.

$$n_o = \frac{f \times 60}{pole - pairs} = \frac{f \times 60}{p/2} \quad \text{rev/min}$$

$$n_o = \frac{f \times 120}{p} \quad \text{rev/min}$$

Where n_o = Synchronous rotational speed in rev/min
         f = Power supply frequency in Hz
         p = Number of motor poles

To establish a current flow in the rotor, there must first be a voltage present across the rotor bars. This voltage is supplied by the magnetic field created by the stator current. The rotating stator magnetic flux, which rotates at *synchronous speed*, passes from the stator iron path, across the air-gap between the stator and rotor and penetrates the rotor iron path as shown in Figure 2.4. As the magnetic field rotates, the lines of flux cut across the rotor conductors. In accordance with *Faraday's Law*, this induces a voltage in the rotor windings, which is dependent on the rate of change of flux.

Since the rotor bars are short circuited by the end-rings, current flows in these bars will set up its own magnetic field. This field interacts with the rotating stator flux to produce the rotational force. In accordance with *Lenz's Law*, the direction of the force is that which tends to reduce the changes in flux field, which means that the rotor will accelerate to follow the direction of the rotating flux.

At starting, while the rotor is stationary, the magnetic flux cuts the rotor at **synchronous speed** and induces the highest rotor voltage and, consequently, the highest rotor current. Once the rotor starts to accelerate in the direction of the rotating field, the rate at which the magnetic flux cuts the rotor windings reduces and the induced rotor voltage decreases proportionally. The frequency of the rotor voltage and current also reduces.

When the speed of the rotor approaches synchronous speed at no load, both the magnitude and frequency of the rotor voltage becomes small. If the rotor reached synchronous speed, the rotor windings would be moving at the same speed as the rotating flux, and the induced voltage (and current) in the rotor would be zero. Without rotor current, there would be no rotor field and consequently no rotor torque. To produce torque, the rotor must rotate at a speed slower (or faster) than the synchronous speed. Consequently, the rotor settles at a speed slightly less than the rotating flux, which provides enough torque to overcome bearing friction and windage. The actual speed of the rotor is called the *slip speed* and the difference in speed is called the *slip*. Consequently, induction motors are often referred to as *asynchronous motors* because the rotor speed is not quite in synchronism with the rotating stator flux. The amount of slip is determined by the *load torque*, which is the torque required to turn the rotor shaft.

For example, on a 4 pole motor, with the rotor running at 1490 r/min on *no-load*, the rotor frequency is 10/1500 of 50 Hz and the induced voltage is approximately 10/1500 of its value at starting. At no-load, the rotor torque associated with this voltage is required to overcome the frictional and windage losses of the motor.

As shaft load torque increases, the slip increases and more flux lines cut the rotor windings, which in turn increases rotor current, which increases the rotor magnetic field and consequently the rotor torque. Typically, the slip varies between about 1% of synchronous speed at no-load to about 6% of synchronous speed at full-load.

$$Slip = s = \frac{(n_0 - n)}{n_0} \, per \, unit \, 1$$

and actual rotational speed is

$$n = n_0(1 - s) \quad rev/min^2$$

Where n_0 = Synchronous rotational speed in rev/min
                 n = Actual rotational speed in rev/min
                 s = Slip in per-unit

The direction of the rotating stator flux depends on the phase sequence of the power supply connected to the stator windings. The phase sequence is the sequence in which the voltage in the 3-phases rises and reaches a peak. Usually the phase sequence is designated A-B-C, L1-L2-L3 or R-W-B (Red-White-Blue). In Europe this is often designated as U-V-W and many IEC style motors use this terminal designation. If two supply connections are changed, the phase sequence A-C-B would result in a reversal of the direction of the rotating stator flux and the direction of the rotor.

2.4   The equivalent circuit

To understand the performance of an AC induction motor operating from a VVVF converter, it is useful to electrically represent the motor by an *equivalent circuit*. This clarifies what happens in the motor when stator voltage and frequency are changed or when the load torque and slip are changed.

There are many different versions of the equivalent circuit, which depend on the level of detail and complexity. The stator current I_S, which is drawn into the stator windings from the AC stator supply voltage V, can then be predicted using this model.

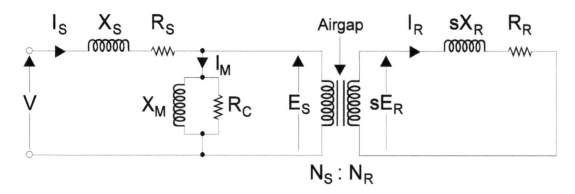

Figure 2.5:
The equivalent circuit of an AC induction motor

Where V = Stator supply voltage R_S = Stator resistance
          E_S = Stator induced voltage X_S = Stator leakage reactance at 50 Hz
          E_R = Rotor induced voltage R_R = Rotor resistance
          N_S = Stator turns X_R = Rotor leakage reactance
          N_R = Rotor turns X_M = Magnetizing inductance
          I_S = Stator current
          I_R = Rotor current
          I_M = Magnetizing current
          R_C = Core losses, bearing friction, windage losses, etc

The main components of the motor electrical *equivalent circuit* are:

- **Resistances** represent the *resistive losses* in an induction motor and comprise,
  - Stator winding resistance losses (R_S)

  - Rotor winding resistance losses (R_R)

  - Iron losses, which depend on the grade and flux density of the core steel

  - Friction and windage losses (R_C)

- **Inductances** represent the *leakage reactance*. These are associated with the fact that not all the flux produced by the stator windings cross the air-gap to link with the rotor windings and not all of the rotor flux enters the air-gap to produce torque.
  - Stator leakage reactance (X_S shown in figure below)

  - Rotor leakage reactance (X_R shown in figure below)

  - Magnetizing inductance (X_M which produces the magnetic field flux)

In contrast with a DC motor, the AC induction motor does not have separate field windings. As shown in the equivalent circuit, the stator current therefore serves a double purpose:

- It carries the current (I_M) which provides the rotating magnetic field
- It carries the current (I_R) which is transferred to the rotor to provide shaft torque

The stator voltage E_S is the *theoretical* stator voltage that differs from the supply voltage by the volt drop across X_S and R_S. X_M represents the magnetizing inductance of the core and R_C represents the energy lost in the core losses, bearing friction and windage losses. The rotor part of the equivalent circuit consists of the induced voltage s.E_R, which as discussed earlier is proportional to the slip and the rotor reactance s.X_R, which depends on frequency and is consequently also dependent on slip.

This equivalent circuit is quite complex to analyze because the transformer, between the stator and rotor, has a ratio that changes when the slip changes. Fortunately, the circuit can be simplified by mathematically adjusting the rotor resistance and reactance values by the turns ratio $N^2 = (N_S/N_R)^2$, i.e. '*transferring*' them to the stator side of the transformer. Once these components have been transferred, the transformer is no longer relevant and it can be removed from the circuit. This mathematical manipulation must also adjust for the variable rotor voltage, which depends on slip. The equivalent circuit can be re-arranged and simplified as shown in the figure below.

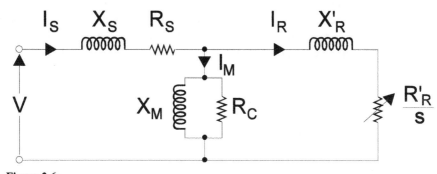

Figure 2.6:
The simplified equivalent circuit of an AC induction motor

Where $\quad X'_R = N^2 \times X_R$
$\qquad\quad R'_R = N^2 \times R_R$
$\qquad\quad N \;\; = N_S/N_R$, the stator/rotor turns ratio

In this modified equivalent circuit, the rotor resistance is represented by an element that is dependent on the slip s. This represents the fact that the induced rotor voltage and consequently current depends on the slip. Consequently, when the induction motor is supplied from a power source of constant voltage and frequency, the current I_S drawn by the motor depends primarily on the slip.

The equivalent circuit can be simplified even further to represent only the most significant components, which are:

- Magnetizing inductance (X_M)
- Variable rotor resistance (R'_R/s)

All other components are assumed to be negligibly small and have been left out.

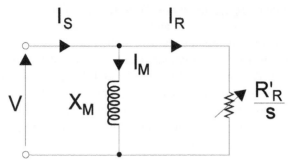

Figure 2.7:
The very simplified equivalent circuit of an AC induction motor

As illustrated above, the total stator current I_S largely represents the vector sum of:

- The **reactive magnetizing current** I_M, which is largely independent of load and generates the rotating magnetic field. This current lags the voltage by 90° and its magnitude depends on the stator voltage and its frequency. To maintain a constant flux in the motor, the V/f ratio should be kept constant.

$$X_M = j\omega L_M = j(2\pi f)L_M$$

$$I_M = \frac{V}{j(2\pi f)\,L_M}$$

$$I_M = k\left(\frac{V}{f}\right) \quad where\ k = constant$$

- The **active current** I_R, which produces the rotor torque depends on the mechanical loading of the machine and is proportional to slip. At no-load, when the slip is small, this current is small. As load increases and slip increases, this current increases in proportion. This current is largely in phase with the stator voltage.

The figure below shows the current vectors for low-load and high-load conditions.

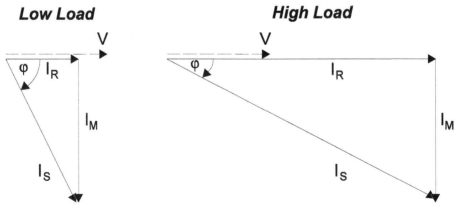

Figure 2.8:
Stator current for low-load and high-load conditions

2.5   **Electrical and mechanical performance**

The angle between the two main stator components of voltage V and current I_S is known as the power factor angle represented by the angle ϕ and can be measured at the stator terminals. As shown, the stator current is the vector sum of the magnetizing current I_M, which is in quadrature to the voltage, and the torque producing current I_R, which is in phase with the voltage. These two currents are not readily available for measurement.

Consequently, the total apparent motor power S also comprises two components, which are in quadrature to one another,

$$S = P + jQ \quad \text{kVA}$$

- *Active power P* can be calculated by

$$P = \sqrt{3} \times V \times I_R \quad \text{kW}$$

or

$$P = \sqrt{3} \times V \times I_S \times \text{Cos}\phi \quad \text{kW}$$

- *Reactive power Q*, can be calculated by

$$Q = \sqrt{3} \times V \times I_M \quad \text{kVAr}$$

or

$$Q = \sqrt{3} \times V \times I_S \times \text{Sin}\phi \quad \text{kVAr}^3$$

Where S = Total apparent power of the motor in kVA
P = Active power of the motor in kW
Q = Reactive power of the motor in kVAr
V = Phase-phase voltage of the power supply in kV
I_S = Stator current of the motor in amps
ϕ = Phase angle between V and I_S (power factor = Cosϕ)

Not all the electrical input power P_I emerges as mechanical output power P_M. A small portion of this power is lost in the stator resistance ($3.I^2.R_S$) and the core losses ($3.I_M^2.R_C$) and the rest crosses the air gap to do work on the rotor. An additional small portion is lost in the rotor ($3I^2R'_R$). The balance is the mechanical output power P_M of the rotor.

Another issue to note is that the magnetizing path of the equivalent circuit is mainly inductive. At no-load, when the slip is small (slip s $\Rightarrow$ 0), the equivalent circuit shows that the effective rotor resistance $R'_R/s \Rightarrow$ infinity. Therefore, the motor will draw only no-load magnetizing current. As the shaft becomes loaded and the slip increases, the magnitude of R'_R/s decreases and the current rises sharply as the output torque and power increases.

This affects the *phase relationship* between the stator voltage and current and the power factor Cosϕ. At no-load, the power factor is low, which reflects the high component of magnetizing current. As mechanical load grows and slip increases, the effective rotor resistance falls, active current increases and power factor improves.

When matching motors to mechanical loads, the two most important considerations are the torque and speed. The *torque–speed curve*, which is the basis of illustrating how the torque changes over a speed range, can be derived from the equivalent circuit and the equations above. By reference to any standard textbook on 3-phase AC induction motors, the *output torque* of the motor can be expressed in terms of the speed as follows:

$$T_M = \frac{3 \times s \times V^2 \times R_R^{'}}{[(R_S + R_R^{'})^2 + s(X_S + X_R^{'})^2] n_o}$$

This equation and the curve in Figure 2.9 (below) shows how the motor output torque T_M varies when the motor runs from standstill to full speed under a constant supply voltage and frequency. The torque requirements of the mechanical load are shown as a dashed line.

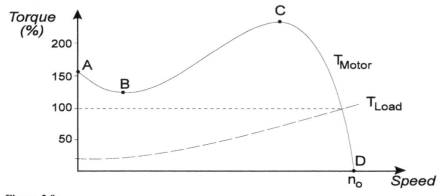

Figure 2.9:
Torque-speed curve for a 3-phase AC induction motor

A: is called the breakaway starting torque

B: is called the pull-up torque

C: is called the pull-out torque (or breakdown torque or maximum torque)

D: is the synchronous speed (zero torque)

At starting, the motor will not pull away unless the starting torque exceeds the load breakaway torque. Thereafter, the motor accelerates if the motor torque always exceeds the load torque. As the speed increases, the motor torque will increase to a maximum T_{Max} at point C.

On the torque–speed curve, the final drive speed (and slip) stabilizes at the point where the *load torque* exactly equals the *motor output torque*. If the load torque increases, the motor speed drops slightly, slip increases, stator current increases, and the motor torque increases to match the load requirements.

The range CD on the torque–speed curve is the stable operating range for the motor. If the load torque increased to a point beyond T_{Max}, the motor would stall because, once the speed drops sufficiently back to the unstable portion ABC of the curve, any increase in load torque requirements T_L and any further reduction in drive speed, results in a lower motor output torque.

The relationship between stator current I_S and speed in an induction motor, at its rated voltage and frequency, is shown in the figure below. When an induction motor is started direct-on-line from its rated voltage supply, the stator current at starting can be as high as 6 to 8 times the rated current of the motor. As the motor approaches its rated speed, the current falls to a value determined by the mechanical load on the motor shaft.

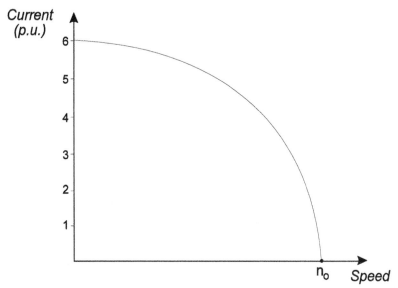

Figure 2.10:
Current–speed characteristic of a 3-phase AC induction motor

Some interesting observations about the AC induction motor that can be deduced from the above equations are:

- Motor output torque is proportional to the square of the voltage

$$T_M \propto V^2$$

Consequently, starting an induction motor with a reduced voltage starter, such as soft-starters, star-delta starters, auto-transformer starters, etc, means that motor starting torque is reduced by the square of the reduced voltage.

- The efficiency of an induction motor is approximately proportional to $(1 - s)$
  i.e. as speed drops and slip increases, efficiency drops

$$Eff \propto (1-s)$$

The induction motor operates as a slipping clutch with the slip power being dissipated as heat from the rotor as 'copper losses'. On speed control systems that rely on slip, such as wound-rotor motors with variable resistors, slip-recovery systems, etc, speed variation is obtained at the cost of motor efficiency.

Efficient use of an induction motor means that slip should be kept as small as possible. This implies that, from an efficiency point of view, the ideal way to control the speed of an induction motor is the stepless control of frequency.

3-phase AC induction motors typically have slip values at full load of,

- 3% to 6% for small motors

- 2% to 4% for larger motors

This means that the speed droop from no-load to full load is small and therefore this type of motor has an almost constant speed characteristic.

One of the most fundamental and useful formulae for rotating machines is the one that relates the **mechanical output power** P_M of the motor to torque and speed,

$$P_M = \frac{(T_M \times n)}{9550} \text{ kW}$$

Where P_M = Motor Output Power in kW
                T_M = Motor Output torque in Nm
                N = Actual Rotational speed in rev/min

2.6   Motor acceleration

An important aspect of correctly matching a motor to a load, is the calculation of the acceleration time of the motor from standstill to full running speed. Acceleration time is important to avoid over-heating the motor due to the high starting currents. So it is often necessary to know how long the machine will take to reach full rated speed. Manufacturers of electric motors usually specify a maximum starting time, during which acceleration can safely take place. This can be a problem during the acceleration of a high inertia load, such as a fan.

Figure 2.9 shows the motor torque curve and the load torque curve plotted on the same graph for a speed range from standstill to full speed. Assuming DOL starting, the time taken to accelerate a mechanical load to full speed depends on:

- **Acceleration torque** (T_A), which is the difference between the motor torque (T_M) and the load torque (T_L), $T_A = (T_M - T_L)$
- **Total moment of inertia** (J_{Tot}) of the rotating parts which is the sum of
  - moment of inertia of the rotor
  - referred value of the moment of inertia of the load

For acceleration to occur, the output torque of the motor must exceed the mechanical load torque. The bigger the *acceleration torque*, the shorter the acceleration time and vice versa. When the motor torque is less than the load torque, the motor will stall. Figure 2.11 shows an example of the *acceleration torque* of a motor, started direct-on-line, driving a centrifugal pump load, whose torque requirement is low at starting and increases as the square of the speed.

The *acceleration torque* at starting is roughly equal to the rated motor torque, increases as the pump drive accelerates and then falls to zero as the motor reaches its rated speed. A steady state speed is reached when *motor torque* T_M matches the *load torque* T_L. The time taken from standstill to reach this stable speed is called the *acceleration time*.

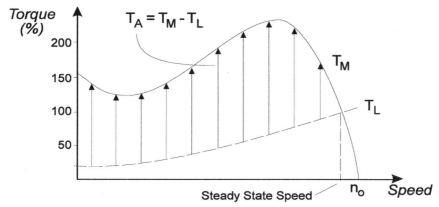

Figure 2.11:
Acceleration torque during the starting of an AC induction motor

The rate of acceleration of the drive system also depends on the *moment of inertia* (*J*) of the rotating object. The higher the value of *J*, the longer it takes to increase speed.

$$T_A = J \frac{d\omega}{dt} \quad Nm$$

Where J = Inertia of the drive system in kgm^2
               ω = Rotational speed in radians/sec

If this formula is adjusted to convert the rotational speed from rad/sec to rev/min

$$T_A = J \frac{2\pi}{60} \frac{dn}{dt} \quad Nm$$

Where n = Rotational speed in rev/min

Re-arranging

$$\frac{dt}{dn} = J \frac{2\pi}{60} \frac{1}{T_A}$$

This is integrated with respect to speed, from starting speed (n_1) to final speed (n_2). The total acceleration time t_d is given by:

$$t_d = J \frac{2\pi}{60} \int_{n_1}^{n_2} \frac{1}{T_A} dn \quad sec$$

If the acceleration torque were constant over the acceleration period, this formula would simplify to:

$$t_d = J \frac{2\pi}{60} \frac{(n_2 - n_1)}{T_A} \quad sec$$

Inertia can be calculated using the formula:

$$J = \frac{G \times D^2}{4} \quad \text{kgm}^2$$

Where J = Moment of inertia of the rotating in kgm^2
G = Mass in kilograms (kg)
D = Diameter of gyration in meters (m)

It is not usually necessary to calculate the value of J from first principles because this may be obtained from the manufacturer of the motor as well as the driven machine.

2.7 AC induction generator performance

The performance of the 3-phase AC induction motor has been described for the speed range from zero up to its rated speed at 50 Hz, where it behaves as a motor. A motor converts electrical energy to mechanical energy. The induction motor will always run at a speed lower than synchronous speed because, even at no-load, a small slip is required to ensure that there is sufficient torque to overcome friction and windage losses.

If, by some external means, the rotor speed was increased to the point that there was no slip, the induced voltage and current in the rotor fall to zero and torque output is zero.

If the rotor speed is, by some external means increased above this, the rotor will run faster than the rotating stator field and the rotor conductors again start to cut the lines of magnetic flux. Induced voltage reappears in the rotor, but in the opposite direction. From Lenz's Law, this results in currents that oppose this change and the power flows in the opposite direction from the driven rotor to the stator. Power flows from the mechanical prime mover, through the induction machine into the electrical supply connected to the stator. When the speed of the machine exceeds the synchronous speed n_o, it then operates as an *induction generator*.

This situation can often occur in the case of cranes, hoists, inclined conveyors, etc, where the load 'over-runs' the motor.

The torque–speed curve can be extended to cover the induction generator region as well. The shape of the curve in the generator region is identical to the motor region because exactly the same equivalent circuit applies. The only difference is that the slip is negative and active power is transferred back into the mains.

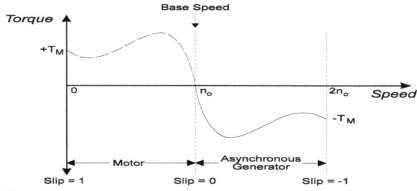

Figure 2.12:
Transition from induction motor to induction generator

2.8   Efficiency of electric motors

The efficiency of a machine is a measure of how well it converts the input electrical energy into mechanical output energy. It is directly related to the losses in the motor, which depend on the design of the machine. Referring to the equivalent circuit of an induction motor, the losses comprise the following:

- Load dependent losses
  These are mainly the copper losses due to the load current flowing through the resistance of the stator and rotor windings and shown in the equivalent circuit as roughly $I_S^2(R_S + R'_R)$. These losses are proportional to the square of the stator current.

- Constant losses
  These losses are mainly due to the friction, windage and iron losses and are almost independent of load. They are represented in the equivalent circuit as $I_M^2 R_C$.

Since the constant losses are essentially independent of load, while the stator and rotor losses depend on the square of the load current, the overall efficiency of an AC induction motor drops significantly at low load levels, as shown in Figure 2.13.

Because of price competition, AC motor manufacturers are under pressure to economize on the quality and quantity of materials used in the motor. Reducing the quantity of copper increases the load dependent losses. Reducing the quantity of iron increases iron losses. Consequently, high efficiency of motors usually cost more. On large motors, high efficiency represents a significant saving in energy costs, which can be offset against the higher initial cost of a more efficient motor.

For electric motors used in AC variable speed drive applications, additional harmonic currents result in additional losses in the motor making it even more desirable to use high efficiency motors.

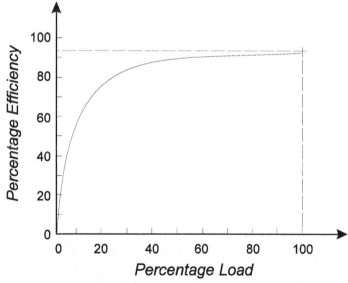

Figure 2.13:
Efficiency of an AC induction motor vs load

2.9   Rating of AC induction motors

AC induction motors should be designed or selected to match the load requirements of any particular application. Some mechanical loads require the motor to run continuously at a particular load torque. Other loads may be cyclical or with numerous stops and starts.

The key consideration in matching a motor to a load is to ensure that the temperature inside the motor windings does not rise, as a result of the load cycle, to a level that exceeds the *critical temperature*. This *critical temperature* is that level which the stator and rotor winding insulation can withstand without permanent damage. Insulation damage can shorten the useful life of the motor and eventually results in electrical faults.

The temperature rise limits of insulation materials are classified by standards organizations, such as IEC 34.1 and AS 1359.32. These standards specify the maximum permissible temperatures that the various classes of insulation materials should be able to withstand. A safe temperature is the sum of the maximum specified ambient temperature and the permitted temperature rise due to the mechanical load.

For purposes of motor design, most motor specifications, such as IEC, AS/NZS, specify a maximum ambient temperature of 40°C. The temperature rise of the induction machine is the permissible increase in temperature, above this maximum ambient, to allow for the losses in the motor when running at full load. The maximum critical temperatures for each insulation class and the temperature rise figures, which are specified by IEC 34.1 and AS 1359.32 for rotating electrical machines, are as follows:

Insulation class	E	B	F	H
Maximum temperature	120°C	130°C	155°C	180°C
Max temperature rise	70°C	80°C	100°C	125°C

Figure 2.14:
Maximum temperature ratings for insulation materials

From these tables, note that electrical rotating machines are designed for an overall temperature rise to a level that is below the maximum specified for the insulation materials.

For example, using class-F insulation,

Max ambient + Max temperature rise = 40°C + 100°C = 140°C

which gives a thermal reserve of 15°C. The larger the thermal reserve, the longer the life expectancy of the insulation material.

When operating continuously at the maximum rated temperature of its class, the life expectancy of the insulation is about 10 years. Most motors do not operate at such extreme conditions because an additional safety margin is usually allowed between the calculated load torque requirements and the actual size of the motor chosen for the application. So life expectancy of a motor, which is correctly matched to its load and with suitable safety margins, can reasonably be taken as between 15 to 20 years.

If additional thermal reserve is required, the motor can be designed for an even lower temperature. It is common practice for the better quality manufacturers to design their motors for class-B temperature rise but to actually use class-F insulating materials. This provides an extra 20°C thermal reserve that will extend the life expectancy to more than

20 years. This also means that the motor could be used at higher ambient temperatures of up to 50°C or more, theoretically up to 65°C.

In manufacturer's catalogues, the rating of 3-phase AC induction motors are usually classified in terms of the following:

- Rated output power, in kW
- Rated speed, depends on the number of poles
- Rated for a continuous duty cycle S1, (see below)
- Rated at an ambient temperature not exceeding 40°C
- Rated at an altitude not exceeding 1000 m above sea level, which implies an atmospheric pressure of above 1050 mbar
- Rated for a relative humidity of less than 90%

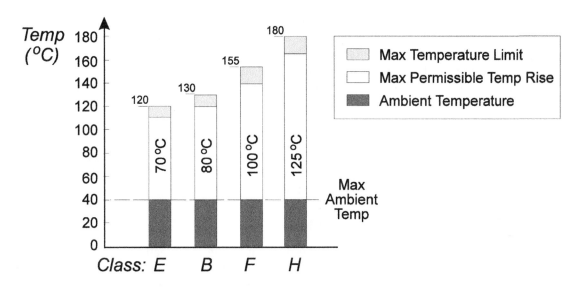

Figure 2.15:
Summary of temperature rise for classes of insulation materials according to IEC 34.1

AC induction motors often need to operate in environmental conditions where the **ambient temperature** and/or the **altitude** exceed the basis for the standard IEC or AS rating.

Where the ambient temperature is excessively high, temperature de-rating tables are available from the motor manufacturers. An example of one manufacturer's de-rating table, for both temperature and altitude, is shown in Figure 2.16 below. As pointed out earlier, better quality AC induction motors have a built-in **thermal reserve**. In some cases, where the ambient temperature is only marginally higher than 40°C, this reserve may be used with no additional de-rating for temperature.

In motor mounting positions that are exposed to continuous direct sunshine, motors should be provided with a protective cover.

Ambient temperature	Permissible output % of rated output	Altitude above Sea Level	Permissible output % of rated output
30°C	107 %	1000m	100 %
40°C	100 %	1500m	96 %
45°C	96 %	2000m	92 %
50°C	92 %	2500m	88 %
55°C	87 %	3000m	84 %
60°C	82 %	3500m	80 %
70°C	65 %	4000m	76 %

Figure 2.16:
Motor de-rating for temperature and altitude

At *high altitudes*, where there is a reduced atmospheric pressure, the cooling of electrical equipment is degraded by the reduced ability of the air to remove the heat from the cooling surfaces of the motor. When the air pressure falls with increased altitude, the density of the air falls and, consequently, its thermal capacity is reduced. In accordance with the standards, AC induction motors are rated for altitudes up to 1000 meters above sea-level. Rated power and torque output should be de-rated for altitudes above that.

When a motor needs to be de-rated for both temperature and altitude, the de-rating factors given in the table above should be multiplied together. For example, for a motor operating at above 2500 m in an ambient temperature of 50°C, the overall de-rating factor should be $(0.92 \times 0.88) \times 100\%$, or 81%.

2.10 Electric motor duty cycles

The rated output of an AC induction motor given in manufacturer's catalogues is based on some assumptions about the proposed application and duty cycle of the motor. It is common practice to base the motor rating on the **continuous running duty cycle** S1. When a motor is to be used for an application duty cycle other than the S1 continuous running duty, some precautions need to be taken in selecting a motor and the standard motors may be re-rated for the application. The duty cycles are normally calculated so that the average load over a period of time is lower than the continuous load rating S1.

In the standards, several different duty cycles are defined. In IEC 34.1 and AS 1359.30, eight different duty types are defined by the symbols S1 to S8 as follows:

S1: Continuous running duty

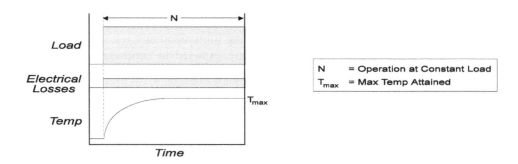

- Operation at constant mechanical load for a period of sufficient duration for thermal equilibrium to be reached.
- In the absence of any indication of the rated duty type of a motor, S1 continuous running duty should be assumed.
- Designation example: S1

S2: Short-time duty

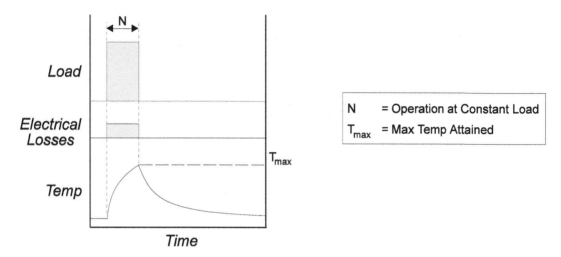

N = Operation at Constant Load
T_{max} = Max Temp Attained

- Operation at constant load, for a period of time which is less than that required to reach thermal equilibrium, followed by a rest and motor de-energised period of sufficient duration for the machine to re-establish temperatures to within 2°C of the ambient or the coolant temperature.
- The values 10 min, 30 min, 60 min and 90 min are recommended periods for the rated duration of the duty cycle.
- Designation example: S2 – 60 min

S3: Intermittent periodic duty not affected by the starting process

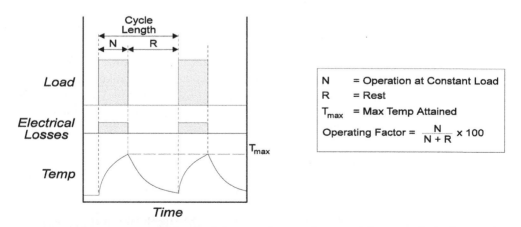

N = Operation at Constant Load
R = Rest
T_{max} = Max Temp Attained
Operating Factor = $\frac{N}{N + R}$ x 100

- A sequence of identical duty cycles, each comprising a period of operation at constant load and a period of rest when the motor is de-energized.

- The period of the duty cycle is too short for thermal equilibrium to be reached.
- Assumed that the starting current does not significantly affect the temperature rise.
- The duration of one duty cycle is 10 min.
- The following items should also be specified for this duty cycle
  - The *cyclic duration factor*, which represents the percentage duration of the loaded period as a percentage of the total cycle
  - Recommended values for cyclic duration factor are 15%, 25%, 40%, 60%
- Designation example: S3 – 25%

S4: Intermittent periodic duty affected by the starting process

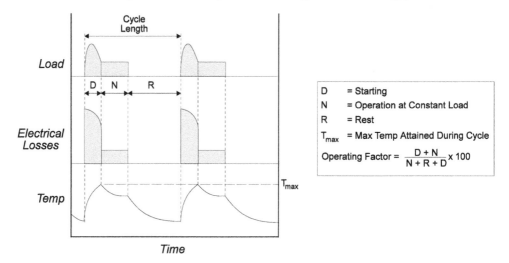

D	= Starting
N	= Operation at Constant Load
R	= Rest
T_{max}	= Max Temp Attained During Cycle
Operating Factor	$= \dfrac{D + N}{N + R + D} \times 100$

- A sequence of identical duty cycles, each comprising a period of significant starting current, a period of operation at constant load and a period of rest when the motor is de-energized.
- The period of the duty cycle is too short for thermal equilibrium to be reached.
- Assumed that the starting current is significant.
- The motor is brought to rest by the load or by mechanical braking, where the motor is not thermally loaded.
- The following items should also be specified for this duty cycle
  - The *cyclic duration factor*, which represents the percentage duration of the loaded period as a percentage of the total cycle
  - The number of *load cycles per hour* (c/h)
  - The *inertia factor* F_I, which is the ratio of the **total** moment of inertia to the moment of inertia of the motor rotor
  - The *moment of inertia* of the motor rotor (J_M)
  - The *average moment of resistance* T_V, during the change of speed given with rated load torque

- Designation example: S4 – 25% – 120 c/h – (F_I = 2) – (J_M = 0.1 kgm^2) – (T_V = 0.5T_N)

S5: Intermittent periodic duty affected by the starting process and also by electric braking

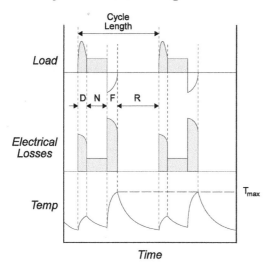

D	= Starting
N	= Operation at Constant Load
F	= Electrical Braking
R	= Rest
T_{max}	= Max Temp Attained During Cycle

$$\text{Operating Factor} = \frac{D + N + F}{D + N + F + R} \times 100$$

- A sequence of identical duty cycles, each comprising a period of significant starting current, a period of operation at constant load, a period of rapid electric braking and a period of rest when the motor is de-energized.
- The period of the duty cycle is too short for thermal equilibrium to be obtained.
- The following items should also be specified for this duty cycle
    - The *cyclic duration factor*, which represents the duration of the loaded period as a percentage of the total cycle
    - The number of *load cycles per hour* (c/h)
    - The *inertia factor F_I*, which is the ratio of the total moment of inertia to the moment of inertia of the motor rotor.
    - The *moment of inertia* of the motor rotor (J_M)
    - The *permissible average moment of resistance T_V*, during the change of speed given with rated load torque.
- Designation example: S5 – 40% – 120 c/h – (F_I = 3) – (J_M = 1.3 kgm^2) – (T_V = 0.3T_N)

S6: Continuous operation, periodic duty with intermittent load

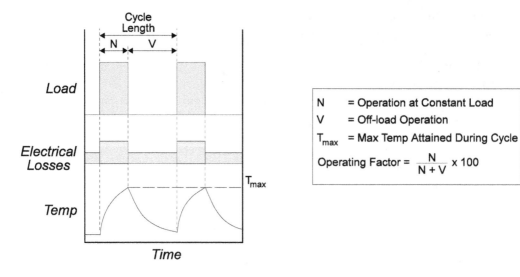

N	= Operation at Constant Load
V	= Off-load Operation
T_{max}	= Max Temp Attained During Cycle

$$\text{Operating Factor} = \frac{N}{N+V} \times 100$$

- A sequence of identical duty cycles, where each cycle consists of a period at constant load and a period of operation at no-load (no-load current only), but with no period of de-energization.
- The period of the duty cycle is too short for thermal equilibrium to be obtained.
- Recommended values for the cyclic duration factor are 15%, 25%, 40% and 60%.
- The duration of the duty cycle is 10 min.
- Designation example: S6 – 40%.

S7: Uninterrupted periodic duty, affected by the starting process and also by electric braking

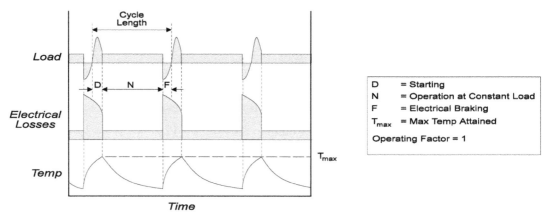

D	= Starting
N	= Operation at Constant Load
F	= Electrical Braking
T_{max}	= Max Temp Attained

Operating Factor = 1

- A sequence of identical duty cycles, each comprising a period of starting current, a period of operation at constant load, a period of electric braking.
- The braking method is too short for thermal equilibrium to be obtained.
- The following items should also be specified for this duty cycle
    - The number of load cycles per hour (c/h)

      – The inertia factor F_I, which is the ratio of the total moment of inertia to the moment of inertia of the motor rotor.

      – The moment of inertia of the motor rotor (J_M)

      – The permissible average moment of resistance T_V, during the change of speed given with rated load torque.

- Designation example: S7 – 500 c/h – ($F_I = 2$) – ($J_M = 0.08$ kgm^2) – ($T_V = 0.3T_N$)

S8: Uninterrupted periodic duty with recurring speed and load changes

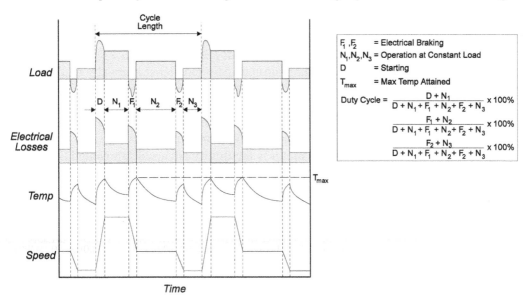

- A sequence of identical duty cycles, each comprising a period of operation at constant load corresponding to a predetermined speed of rotation, followed by one or more periods of operation at other constant loads corresponding to different speeds of rotation.
- The period of the duty cycle is too short for thermal equilibrium to be obtained.
- This type of duty cycle is used for pole changing motors.
- The following items should also be specified for this duty cycle
  - The number of load cycles per hour (c/h).

  - The inertia factor F_I, which is the ratio of the total moment of inertia to the moment of inertia of the motor rotor.

  - The permissible average moment of resistance T_V, during the change of speed given with rated load torque.

  - The cyclic duration factor for each speed of rotation.

  - The moment of inertia of the motor rotor (J_M).

  - The combinations of the load and the speed of rotation are listed in the order in which they occur in use.

- Designation examples:
  - $S8 - 30 \text{ c/h} - (F_I = 30) - T_V = 0.5T_N - 24 \text{ kW} - 740 \text{ rev/m} - 30\%$
  - $S8 - 30 \text{ c/h} - (F_I = 30) - T_V = 0.5T_N - 60 \text{ kW} - 1460 \text{ rev/m} - 30\%$
  - $S8 - 30 \text{ c/h} - (F_I = 30) - T_V = 0.5T_N - 45 \text{ kW} - 980 \text{ rev/m} - 40\% - (J_M = 2.2 \text{ kgm}^2)$

2.11 Cooling and ventilation of electric motors (IC)

All rotating electrical machines generate heat as a result of the electrical and mechanical losses inside the machine. Losses are high during starting or dynamic braking. Also, losses usually increase with increased loading. Cooling is necessary to continuously transfer the heat to a cooling medium, such as the air. The different methods of cooling rotating machines are classified in the standards IEC 34.6 and AS 1359.21.

For AC induction motors, cooling air is usually circulated internally and externally by one or more fans mounted on the rotor shaft. To allow for operation of the machine in either direction of rotation, fans are usually of the bi-directional type and made of a strong plastic material, aluminum, or steel. In addition, the external frames of the motor are usually provided with cooling ribs to increase the surface area for heat radiation.

The most common type of AC motor is the **totally enclosed fan cooled** (TEFC) motor, which is provided with an external forced cooling fan mounted on the non-drive end (NDE) of the shaft, with cooling ribs running axially along the outer surface of the motor frame. These are designed to keep the air flow close to the surface of the motor along its entire length, thus improving the cooling and self-cleaning of the ribs. An air-gap is usually left between the ribs and the fan cover for this purpose.

Internally, on smaller TEFC motors, the rotor end-rings are usually constructed with ribs to provide additional agitation of the internal air for even distribution of temperature and to allow the radiation of heat from the end shields and frame.

Special precautions need to be taken when standard TEFC induction motors are used with AC variable speed drives, powered by VVVF converters. For operation at speeds below the rated frequency of 50 Hz, the shaft mounted fan cooling efficiency is lost. For constant torque loads, it is sometimes necessary to install a separately powered forced cooling fan (IC 43) to maintain adequate cooling at low speeds. On the other hand, for prolonged operation at high speeds above 50 Hz, the shaft mounted fan works well but may make excessive noise. Again, it may be advisable to fit a separately powered cooling fan.

Larger rotating machines can have more elaborate cooling systems with heat exchangers.

The system used to describe the method of cooling is currently being changed by IEC, but the designation system currently in use is as follows:

- A prefix comprising the letters IC (index of cooling)
- A letter designating the cooling medium, this is omitted if only air is used
- Two numerals which represent
  1. The cooling circuit layout

  2. The way in which the power is supplied to the circulation of the cooling fluid, fan, no fan, separate forced ventilation, etc

Code	Description	Drawing
IC 01	- Open machine - Fan mounted on shaft - Often called 'drip-proof' motor	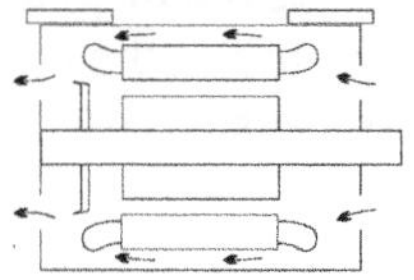
IC 40 (New : IC 410)	- Enclosed machine - Surface cooled by natural convection and radiation - No external fan	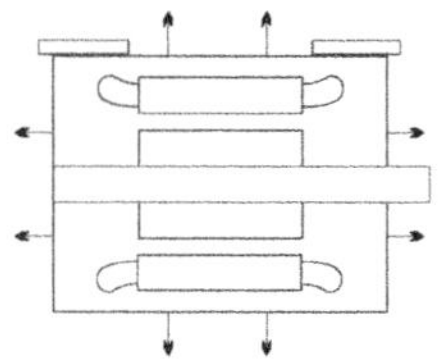
IC 41 (New : IC 411)	- Enclosed machine - Smooth or finned casing - External shaft-mounted fan - Often called TEFC motor	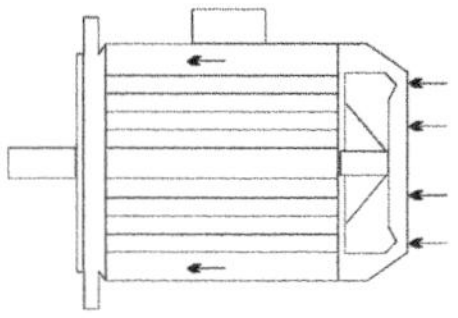
IC 43 A (New : IC 416A)	- Enclosed machine - Smooth or finned casing - External motorized Axial fan supplied with machine	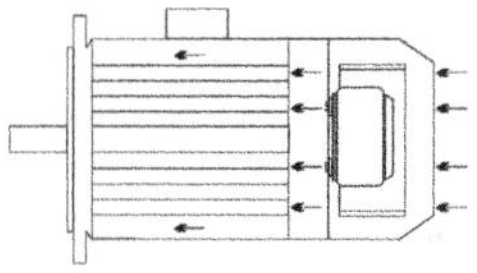
IC 43 R (New : IC 416R)	- Enclosed machine - Smooth or finned casing - External motorized Radial fan supplied with machine	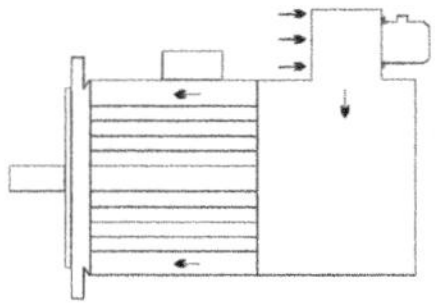
IC 61 (New : IC 610)	- Enclosed machine - Heat Exchanger fitted - Two separate air circuits - Shaft-mounted Fans - Often called CacA motor	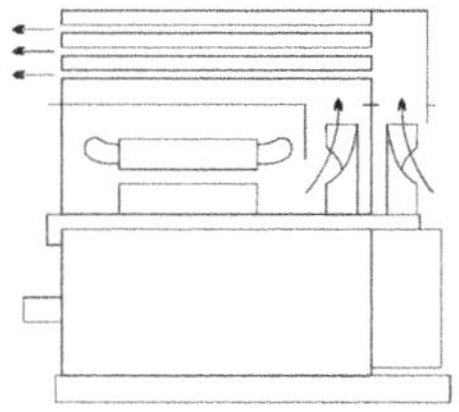

Figure 2.17:
Designation of the most common methods of cooling

2.12 Degree of protection of motor enclosures (IP)

The **degree of protection** (also called *index of protection – IP*) which is provided by the enclosure of the motor, is classified in the standards IEC 34.5 and AS 1359.20.

The system used to describe the Index of Protection is as follows:

- A prefix comprising the letters IP (index of protection)
- Three numerals which represent
  1. The protection against contact and ingress of solid objects, such as dust.
  2. The protection against ingress of liquids, such as water.
  3. The mechanical protection and its resistance to impact.

This third numeral is often not used in practice.

First number : protection against solid objects			Second number : protection against liquids			Third number : mechanical protection		
IP	Tests	Definition	IP	Tests	Definition	IP	Tests	Definition
0		No protection	0		No protection	0		No protection
1	Ø 50 mm	Protected against solid objects of over 50 mm (eg : accidental hand contact)	1		Protected against vertically dripping water (condensation)	1	150g 15cm	Impact energy : 0.225 J
2	Ø 12 mm	Protected against solid objects of over 12 mm (eg : finger)	2	15°	Protected against water dripping up to 15° from the vertical	2	250g 15cm	Impact energy : 0.375 J
3	Ø 2.5 mm	Protected against solid objects of over 2.5 mm (eg : tools, wire)	3	60°	Protected against rain falling up to 60° from the vertical	3	250g 20cm	Impact energy : 0.500 J
4	Ø 1 mm	Protected against solid objects of over 1 mm (eg : small tools, thin wire)	4		Protected against water splashes from all directions			
5		Protected against dust (no deposits of harmful material)	5		Protected against jets of water from all directions	5	500g 40cm	Impact energy : 2 J
6		Totally protected against dust. Does not involve rotating machines	6		Protected against jets of water comparable to heavy seas			
			7	0.15m	Protected against the effects of immersion to depths of between 0.15 and 1 m	7	1500g 40cm	Impact energy : 6 J
			8	m	Protected against the effects of prolonged immersion at depth			
						9	5000g 40cm	Impact energy : 20 J

Figure 2.18:
Summary of the index of protection

This system of degrees of protection does not relate to protection against corrosion.

For example, a machine with an index of protection of **IP557**, is protected as follows:

5: Machine is protected against accidental personal contact of moving parts, such as the fan, and against the ingress of dust

Test Result: No risk of direct contact with rotating parts (test finger)

No risk that dust could enter the machine in harmful quantities

5: Machine is protected against jets of water from all directions from hoses 3 m away and with a flow rate less than 12.5 liters/sec at 0.3 bar

Test Result: No damage from water projected onto the machine during operation

7: Machine is resistant to impacts of up to 6 joules

Test Result: Damage caused by impacts does not affect running of the machine

2.13 Construction and mounting of AC induction motors

Modern squirrel cage AC induction motors are available in several standard types of construction and mounting arrangements. These are classified in accordance with the standards IEC 34.7 and AS 1359.22.

Mounting position needs to be specified to ensure that drain plugs, bearings and other mechanical details are correctly located and dimensioned during assembly.

The system used to describe the mounting arrangements is as follows:

- A prefix comprising the letters IM (index of mounting)
- Four numerals which represent,
  1. Type of construction

  2. Type of construction

  3. Mounting Position

  4. Mounting Position

A summary of the mounting designations is shown in Figure 2.19.

A previous system of designation used letters *B* (horizontal mounting) and *V* (vertical mounting). This system has been superseded in both IEC 34.7 and AS 1359.22. The old designations are shown in the table in brackets.

Foot Mounted Motors			
IM 1001 (IM B3) - Horizontal shaft - Feet on floor	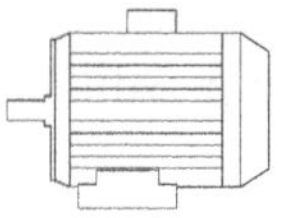	**IM 1071 (IM B8)** - Horizontal shaft - Feet on ceiling	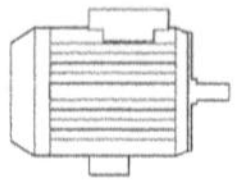
IM 1051 (IM B6) - Horizontal shaft - Feet wall mounted with feet on LHS when viewed from drive end	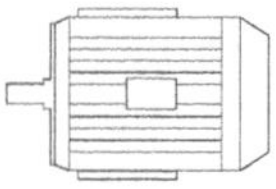	**IM 1011 (IM V5)** - Vertical shaft - Shaft facing down - Feet on wall	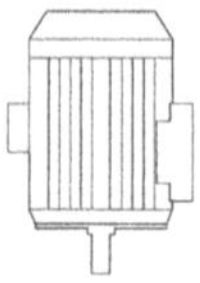
IM 1061 (IM B7) - Horizontal shaft - Feet wall mounted with feet on RHS when viewed from drive end	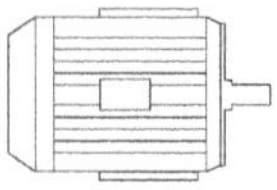	**IM 1031 (IM V6)** - Vertical shaft - Shaft facing up - Feet on wall	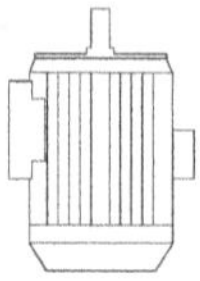

Figure 2.19:
Mounting designations for foot mounted motors

Flange Mounted Motors			
IM 3001 (IM B5) - Horizontal shaft	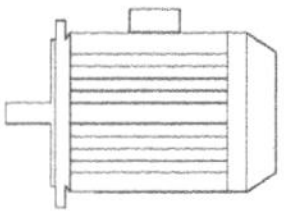	**IM 2001 (IM B35)** - Horizontal shaft - Feet on floor	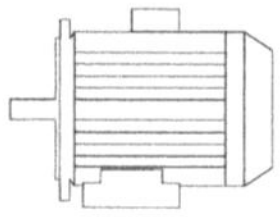
IM 3011 (IM V1) - Vertical shaft - Shaft facing down	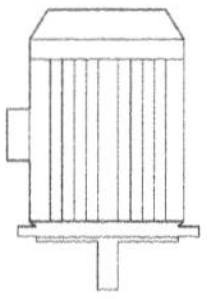	**IM 2011 (IM V15)** - Vertical shaft - Shaft facing down - Feet on wall	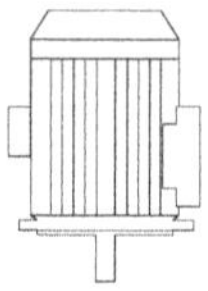
IM 3031 (IM V3) - Vertical shaft - Shaft facing up	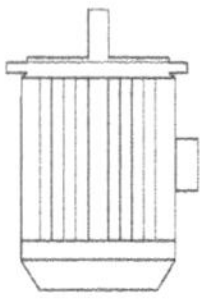	**IM 2031 (IM V36)** - Vertical shaft - Shaft facing up - Feet on wall	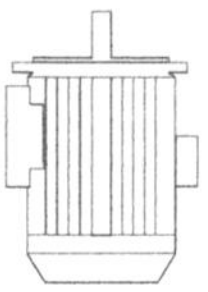

Figure 2.20:
Mounting designations for flange mounted motors

Face Mounted Motors			
IM 3601 (IM B14) - Horizontal shaft	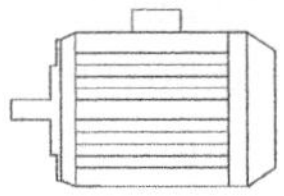	**IM 2101 (IM B34)** - Horizontal shaft - Feet on floor	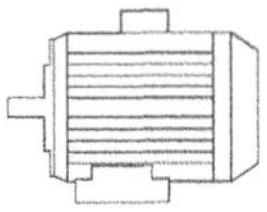
IM 3611 (IM V18) - Vertical shaft - Shaft facing down	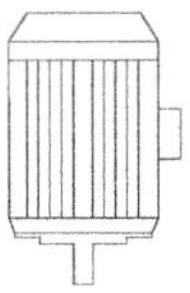	**IM 2111 (IM V58)** - Vertical shaft - Shaft facing down - Feet on wall	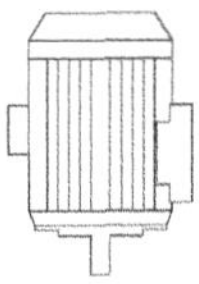
IM 3631 (IM V19) - Vertical shaft - Shaft facing up	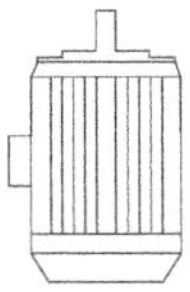	**IM 2131 (IM V69)** - Vertical shaft - Shaft facing up - Feet on wall	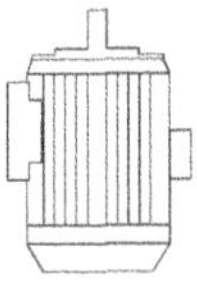

Figure 2.21:
Mounting designations for face mounted motors

2.14   Anti-condensation heaters

When rotating electrical machines need to stand idle for long periods of time in severe climatic conditions, such as a high humidity environment, moisture can be drawn into the machine and absorbed into and onto the insulation of the stator and rotor windings. When a machine is de-energized after it has been running for a period of time, the internal temperature is high. As the machine cools, the low pressure inside the machine draws external moist air into the machine via the seals around the shaft. The moisture degrades the performance of the insulation materials by providing a partially conductive path between the windings and the frame of the machine. When the machine is energized, electrical breakdown of the insulation can occur. Standby motors or generators, which have not been used for some time, can fail to operate when they are needed.

Under these conditions, where a motor is expected to stand idle for long periods in an environment of high humidity, it may be necessary to specify additional winding impregnation treatment and consideration should also be given to anti-condensation heaters. These are fitted inside the motor and their connections brought out to terminals. The heaters are energized from a 240 V supply when the motor is not in use to prevent condensation forming inside the windings.

Anti-condensation heaters are normally in the form of a tape, which comprises a flat glass-fiber tape with a heating element woven into it. This tape is then inserted inside a glass fiber sleeve and wrapped around the stator winding overhang, braced and impregnated with the stator winding. One heater element is normally fitted to each end of the stator winding. A typical rating of a heater varies from 25 watts, for small motors, to 200 watts for large motors.

2.15 Methods of starting AC induction motors

Direct-on-line (DOL) starting is the simplest and most economical method of starting an AC squirrel cage induction motor. A suitably rated contactor is used to connect the stator windings of the motor directly to the 3-phase power supply. While this method is simple and produces a reasonable level of starting torque, there are a number of disadvantages:

- The starting current is very high, between 3 to 8 times the full load current. Depending on the size of the motor, this can result in voltage sags in the power system.
- The full torque is applied instantly at starting and the mechanical shock can eventually damage the drive system, particularly with materials handling equipment, such as conveyors.
- In spite of the high starting current, for some applications the starting torque may be relatively low, only 1.0 to 2.5 times full load torque.

To overcome these problems, other methods of starting are often used. Some common examples are as follows:

- Star-delta starting
- Series inductance starting (e.g. series chokes)
- Auto-transformer starting
- Series resistance starting (e.g. liquid resistance starter)
- Solid state soft-starting (e.g. smart motor controller)
- Rotor resistance starting, requires a slipring motor

Most of the above motor starting techniques *reduce the voltage* at the motor stator terminals, which effectively reduces the starting current as well as the starting torque.

From the equivalent circuits and formulae for AC induction motors, covered earlier in this chapter, the following conclusions can be drawn about *reduced voltage starting*:

- Both the stator current and output torque during starting are proportional to the square of the voltage. During star-delta starting, the voltage is reduced to 0.58 of its rated value. The current and torque are reduced to 0.33 of prospective value.

$$I_{Start} \propto (Voltage)^2$$

$$T_{Start} \propto (Voltage)^2$$

2.16 Motor selection

The correct selection of an AC induction motor is based on a thorough understanding of the application for which the motor is to be used. This requires knowledge about the type and size of the mechanical load, its starting and acceleration requirements, running speed requirements, duty cycle, stopping requirements, and the environmental conditions. The following checklist and reference to the preceding sections provides a guide to the selection procedure.

When selecting an electric motor, the following factors should be considered:

- Type and torque requirements of the mechanical load
- Method of starting

- Acceleration time
- Type of construction of AC induction motor
    - squirrel cage rotor
    - wound rotor with sliprings
    - foot mounted
    - flange mounted
- Environmental conditions
    - ambient temperature
    - altitude
    - dust conditions
    - water
- Required degree of protection of the enclosure
- Insulation class
- Motor protection
- Method of cooling
- Mounting arrangement
    - horizontal
    - vertical
- Cable connections
- Direction of rotation
- Duty cycle
- Speed control (if required)

In general, the selection of the motor is dictated by the type of load and the environment in which it will operate. The selection of a cage motor or slipring motor is closely related to the size of the machine, the acceleration time required (determined by load) and the method of starting (determined by the electrical supply limitations).

From the point of view of price, reliability, and maintenance, the cage motor is usually the first choice. In general, slipring motors are required when:

- The load has a high starting torque requirement, but the supply dictates a low starting current
- The acceleration time is long due to high load inertia, such as a fan
- Where duty dictates frequent starting, inching or plugging

These are general comments because cage motors can be successfully used in all the above situations.

Slipring motors are sometimes used for limited speed control. The slip can be controlled by controlling the external rotor resistance. As demonstrated earlier, the overall efficiency of this method is poor, so this method can only be used if the speed does not deviate too far from the rated speed. The slip power is dissipated as heat in the external rotor resistors.

3

Power electronic converters

3.1 Introduction

This chapter deals with the active components (e.g. diodes, thyristors, transistors, etc) and passive components (e.g. resistors, chokes, capacitors, etc) used in *power electronic* circuits and converters. *Power electronics* is that field of electronics which covers the conversion of electrical energy from one form to another for high power applications. It applies to circuits in the following power ranges:

- Power ratings up to the MVA range
- Frequency ratings up to about 100 kHz

Power electronics is a rapidly expanding field in electrical engineering and the scope of the technology covers a wide spectrum. Therefore, the emphasis will be on the components used in converters used for the speed control of electric motors. Components used for other applications such as power supplies, high frequency generators, etc will not be covered in great detail.

3.2 Definitions

The following are the common terms used in the field of power electronics.

- **Power electronic components**, are those semiconductor devices, such as diodes, thyristors, transistors, etc that are used in the power circuit of a converter. In power electronics, they are used in the non-linear switching mode (on/off mode) and not as linear amplifiers.
- **Power electronic converter** or 'converter' for short, is an assembly of power electronic components that converts one or more of the characteristics of an electric power system. For example, a converter can be used to change
  - AC to DC

- DC to AC

- Frequency

- Voltage level

- Current level

- Number of phases

The following graphic symbols are used to designate the different types of converter.

- **Rectifier** is that special type of converter that converts AC to DC

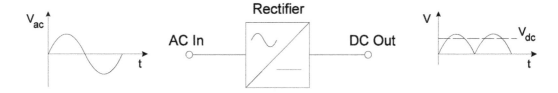

- **Inverter** is that special type of converter that converts DC to AC.

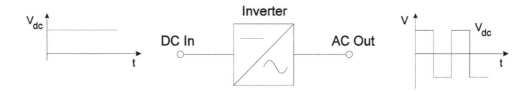

- **AC converter** is that special type of converter that converts AC, of one voltage and frequency, to AC of another voltage and frequency, which are often variable.
An *AC frequency converter* is a special type of AC converter.

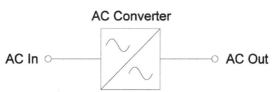

In a power electronic AC converter, it is common to use an intermediary DC link with some form of smoothing.

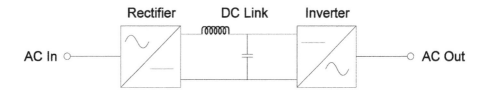

- **DC converter** is one that converts DC of one voltage to DC of another voltage.

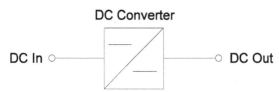

DC Converter

DC In ○――――――――○ DC Out

In a DC converter, it is common to use an intermediary AC link, usually with galvanic isolation via a transformer.

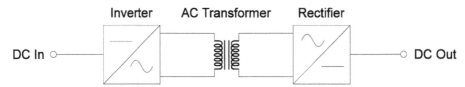

Inverter   AC Transformer   Rectifier

DC In ○――――――――――――――○ DC Out

- **Electronic switch** is one that electronically connects or disconnects an AC or DC circuit and can usually be switched ON and/or OFF. Conduction is usually permitted in **one direction only.**

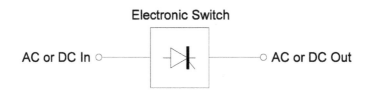

Electronic Switch

AC or DC In ○――――――――○ AC or DC Out

The following components are those devices that are most commonly used as electronic switches in *power electronic converters*. Developments in semiconductor technology have made these power electronic components smaller, more reliable, more efficient (lower losses), cheaper and able to operate at much higher voltages, currents and frequencies. The idealised operating principles of these components can be described in terms of simple mathematical expressions.

- Power diodes
- Power thyristors
- Gate turn-off thyristors (GTO)
- MOS controlled thyristors (MCT)
- Power bipolar junction transistors (BJT)
- Field effect transistors (FET, MOSFET)
- Integrated gate bipolar transistors (IGBT)
- Resistors (provide resistance)
- Reactors or chokes (provide inductance)
- Capacitors (provide capacitance)

In power electronic circuits, semiconductor devices are usually operated in the bi-stable mode, which means that they are operated in either one of two stable conditions:

- Blocking mode: fully switched OFF
    - Voltage across the component is **high**

    - Current through the component is **low** (only leakage current)

- Conducting Mode: fully switched ON
    - Voltage across the component is **low**
    - Current through the component is **high**

Diodes and thyristors are inherently bi-stable but transistors are not. Transistors must be biased fully ON to behave like bi-stable devices.

3.3   Power diodes

A power diode is 2-terminal semiconductor device with a relatively large single P-N junction. It consists of a 2-layer silicon wafer attached to a substantial copper base. The base acts as a heat-sink, a support for the enclosure and also one of the electrical terminals of the diode. The other surface of the wafer is connected to the other electrical terminal. The enclosure seals the silicon wafer from the atmosphere and provides adequate insulation between the two terminals of the diode. The two terminals of a diode are called the *anode (A)* and the *cathode (K)*. These names are derived from the days when *Valves* were commonly used.

SYMBOL:

IDEAL:
   Forward conduction: Resistanceless
   Reverse blocking: Lossless
   Switch on/off time: Instantaneous

Many different mechanical designs are commonly used for diodes, some of which are shown below. Power diodes rated from a few amperes are usually stud mounted but it is increasingly common (more economical) to have several diodes encapsulated into an insulated module. Examples are full wave rectifiers, 6-pulse diode bridges, etc.

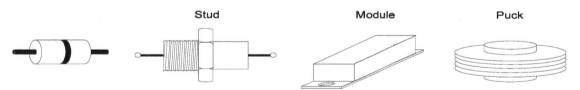

Figure 3.1:
Typical mechanical construction of diodes

The base of this type of diode module is usually not electrically active, so it can be mounted directly onto the heat-sink of a converter. Larger units for high current ratings are usually of the disc type, which provides a larger area of contact between the case and the heat-sink for better cooling.

When the anode is positive relative to the cathode, it is said to be *forward biased* and the diode conducts current. When the anode is negative relative to the cathode the diode is said to be *reverse biased* and the flow of current is blocked. The typical characteristic of a power diode is shown in the figure below.

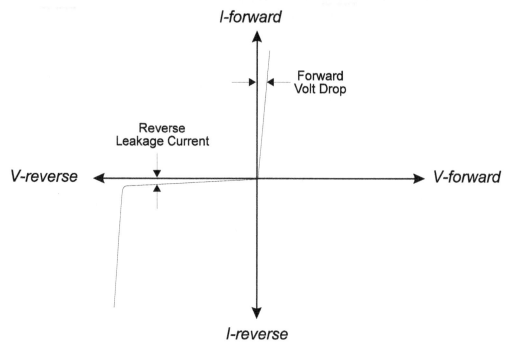

Figure 3.2:
Typical characteristic of a power diode

Unfortunately, power diodes have several limitations:

- In the *conduction mode*, when the diode is *forward biased*
  - Real diodes are not resistanceless and there is a *forward volt drop* of between 0.5 to 1.0 volts during conduction

  - As a result, there is a limit to how much current can continuously flow without overheating. This is the maximum rated current of the diode.

- In the blocking mode, when the diode is *reverse biased*
  - there is a small leakage current

  - there is a limit to how much voltage it can withstand before reverse breakdown and current can start to flow in the reverse direction. It is sound common practice to select diodes with a reverse voltage limit of at least twice the value that will practically occur.

- The *commutation time* from the blocking mode to the conduction mode and vice versa takes a finite time.

A power diode must be rated for the electrical environment in which it is to be used. The following are the most important factors that must be considered when choosing a power diode for a converter application:

- **Forward current rating**. The current rating is based on a certain wave shape and should be taken as a guide only. The real selection should be based on the total power losses in the diode taking into account the actual wave shape, load cycle and cooling conditions.

- **Forward voltage drop**. This has an effect on current sharing between parallel circuits that include diodes.
- **Forward surge current** capability (rate of rise of current di/dt)
- **Reverse voltage rating** (sometimes referred to as PIV - peak inverse voltage)
- **Reverse recovery current di/dt**. This should be taken into account when considering the commutation transients in the diode circuit.
- **I^2t rating**. This is a measure of the energy that a diode can handle in the case of a short circuit without permanent damage. It gives a guide to the correct choice of high speed fuses to protect the diode. Briefly, a protection fuse must be chosen with an I^2t rating lower than the diode.

Depending on the application requirements, various types of diode are available:

- Schottky diodes
- These diodes are used where a low forward voltage drop, typically 0.4 volts, is needed for low output voltage circuits. These diodes have a limited blocking voltage capability of 50 to 100 volts.
- Fast recovery diodes
- These diodes are designed for use in circuits where fast recovery times are needed, for example in combination with controllable switches in high frequency circuits. Such diodes have a recovery time (t_{RR}) of less than a few microsecs.
- Line frequency diodes
- The on-state voltage of these diodes is designed to be as low as possible to ensure that they switch on quickly in rectifier bridge applications. Unfortunately the recovery time (t_{RR}) is fairly long, but this is acceptable for line-frequency rectifier applications. These diodes are available with blocking voltage ratings of several kV and current ratings of several hundred kamps. In addition, they an be connected in series or parallel to satisfy high voltage or current requirements.

3.4 Power thyristors

Thyristors are often referred to as SCRs (silicon controlled rectifiers). This was the name originally given to the device when it was invented by General Electric (USA) in about 1957. This name has never been universally accepted and used. The name accepted by both the IEC and ANSI/IEEE is *reverse blocking triode thyristor* or simply *thyristor*. The name *thyristor* is a generic term that is applied to a family of semiconductor devices that have the regenerative switching characteristics. There are many devices in the thyristor family including the power thyristor, the gate turn-off thyristor (GTO), the field controlled thyristor (FCT), the triac, etc.

A thyristor consists of a 4-layer silicon wafer with 3 P-N junctions. It has two power terminals, called the *anode (A)* and *cathode (K)*, and a third control terminal called the *gate (G)*. High voltage, high power thyristors sometimes also have a 4th terminal, called an auxiliary cathode and used for connection to the triggering circuit. This prevents the main circuit from interfering with the gate circuit.

A thyristor is very similar to a power diode in both physical appearance and construction, except for the gate terminal required to trigger the thyristor into the conduction mode.

SYMBOL:

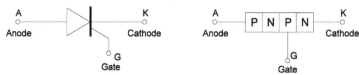

IDEAL: Forward conduction: Resistanceless
Forward blocking: Lossless (no leakage current)
Reverse blocking: Lossless (no leakage current)
Switch on/off time: Instantaneous

As with power diodes, smaller units are usually of the stud type but it is also increasingly common to have 2 or more thyristors assembled into a thyristor module. The base of this type of pack is not electrically active, so it can be mounted directly onto the heat-sink of a converter. Large thyristor units are usually of the disc type for better cooling.

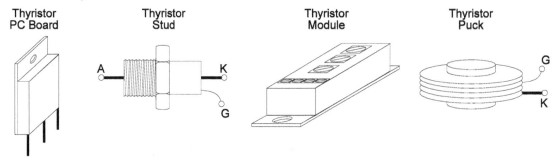

Figure 3.3:
Typical mechanical construction of thyristors

Most converters for the speed control of motors are air-cooled, the smaller units using natural convectional cooling over the heat-sink and the larger units using a fan for forced cooling.

A *thyristor is a controllable device*, which can be switched from a blocking state (high voltage, low current) to a conducting state (low voltage, high current) by a suitable gate pulse. Forward conduction is blocked until an external positive pulse is applied to the gate terminal. A thyristor cannot be turned off from the gate. During forward conduction, its behavior resembles that of a power diode and it also exhibits a forward voltage drop of between 1 to 3 volts. Like the diode, conduction is blocked in the reverse biased direction. A typical characteristic of the thyristor is shown in the Figure 3.4.

There are several ways in which a thyristor can be turned on or brought into *forward conduction*.

- **Positive current gate pulse**. This is the normal way that a thyristor is brought into conduction. The gate pulse must be of a suitable amplitude and duration, depending on the size of the thyristor.
- **High forward voltage**. An excessively high forward voltage between the anode and the cathode can cause enough leakage current to flow to trigger the turn on process.
- **High rate of rise of forward voltage**, dV/dt. A high dV/dt can produce enough leakage current to trigger the turn on process.

- **Excessive temperature**. The leakage current increases with temperature, so high temperature can aggravate the above two problems.

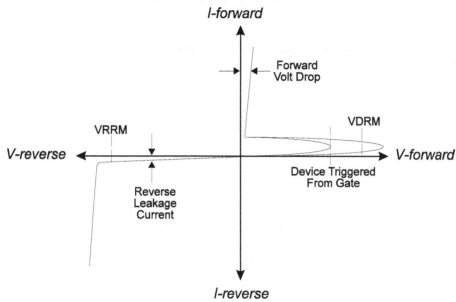

Figure 3.4:
Typical characteristic of a thyristor

A thyristor must be suitable for the electrical environment in which it is used. The following are some of the more important factors which must be considered when choosing a thyristor for a converter application:

- Same factors outlined above for diodes.
- The **power losses** in the thyristor comprise the conduction losses, switching losses (turn on and turn off), gate power losses, forward off state losses and reverse blocking losses. The data sheet usually provides curves for estimating power losses for various wave shapes.
- **Peak forward voltage** (PFV). This is the forward anode voltage that the device must withstand without switching on and without damage.
- **Rate of rise of forward voltage** dV/dt should not be too high, typically it should be less than about 200 Volt/μsec. A parallel RC snubber circuit is usually required to protect the thyristor.
- **Rate of rise of anode current** di/dt should not be too high, typically it should be less than about 100 amp/μsec. The current is initially concentrated around the gate and takes a finite time to spread over the conducting area.

If the rate of rise is too high, local overheating could damage the thyristor. Circuit inductance is usually required to limit the rate of rise of current.

- **Holding current**. The minimum forward current required for the thyristor to maintain forward conduction.
- **Latching current**. The minimum forward current that causes the thyristor to initially latch. This is usually higher than the holding current and is important because the gate pulse may be relatively short.

- **Gate triggering** requirements. A relatively small gate pulse will turn the thyristor on. Typically a value of 100 mA for 10 μsec is the threshold. In practice, a much higher value should be used for optimum thyristor operation. Also, the turn on time is affected by the magnitude of the gate pulse.

The thyristor is turned off when it becomes reverse biased and/or the forward current falls below the holding current. This must be controlled externally in the power circuit.

3.5 Commutation

The transitional period from blocking to conducting, and vice versa, is called *commutation* and the period during which a component turns on/off, is called the *commutation period*. During commutation, the component comes under electrical stress due to changes in the circuit conditions and the thermal stress due to losses. These losses produce heat in the component and also stress the insulation and current paths.

- In the *blocking mode*, losses are usually small and mainly due to the leakage current flowing through the device
- In the *conducting mode*, losses are relatively higher and mainly due to the current and forward volt drop across the component (I^2R losses)
- During *commutation*, losses are due to the transitional voltage and current activity within the component and in the control circuit to trigger the gate.

Figure 3.5 illustrates thyristor commutation for both the turn-on and the turn-off periods.

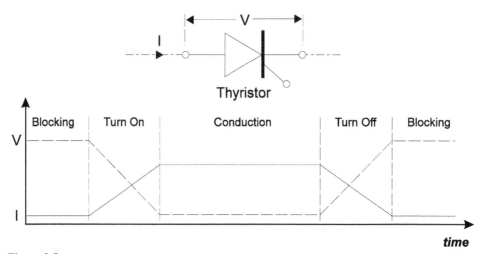

Figure 3.5:
Simple commutation of an electronic switch

In modern PWM inverters, there is a tendency to use electronic switches operating at high switching frequencies to achieve faster responses or better output wave-shapes. Unfortunately, the increased number of commutations results in higher losses both in the triggering circuits as well as the power circuits of the components.

Losses may be reduced by using devices that have the following characteristics:

- Low leakage current during blocking

- Low forward volt drop during conduction
- High switching speed, short commutation period
- Low triggering losses in the control circuit

3.6   Power electronic rectifiers (AC/DC converters)

The first stage of an AC frequency converter is the conversion of a 3-phase AC power supply to a smooth DC voltage and current. Simple bi-stable devices, such as the diode and thyristor, can effectively be used for this purpose.

Initially, when analyzing power electronic circuits, it will be assumed that the bi-stable semiconductor devices, such as the diodes and thyristors, are ideal switches with no losses and minimal forward voltage drop. It will also be assumed that the reactors, capacitors, resistors, and other components of the circuits have ideal linear characteristics with no losses. Once the operation of a circuit is understood, the imperfections associated with the practical components can be introduced to modify the performance of the power electronic circuit.

In power electronics, the operation of any converter is dependent on the *switches* being turned ON and OFF in a sequence. Current passes through a switch when it is ON and is blocked when it is OFF. As mentioned above, the word *commutation* is used to describe the transfer of current from one switch turning OFF to another turning ON.

In a diode rectifier circuit, a diode turns ON and starts to conduct current when there is a forward voltage across it, i.e. the forward voltage across it becomes positive. This process usually results in the forward voltage across another diode becoming negative, which then turns off which stops conducting current. In a thyristor rectifier circuit, the switches additionally need a gate signal to turn them on and off.

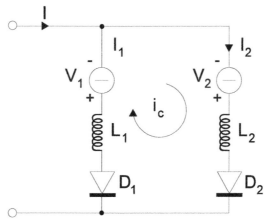

Figure 3.6:
Simple circuit to illustrate commutation from diode D1 to D2

The factors affecting commutation may be illustrated in the idealized diode circuit in Figure 3.6, which shows two circuit branches, each with its own variable DC voltage source and circuit inductance. Assume, initially, that a current I is flowing through the circuit and that the magnitude of the voltage V_1 is larger than V_2. Since $V_1 > V_2$, diode $D1$ has a positive forward voltage across it and it conducts a current I_1 through its circuit inductance L_1. Diode D_2 has a negative forward voltage across it and is blocking and carries no current.

Consequently, at time t_1

$$I_1 = I$$

$$I_2 = 0$$

Suppose that voltage V_2 is increased to a value larger than V_1, the forward voltage across diode D_2 becomes positive and it then starts to turn on. However, the circuit inductance L_1 prevents the current I_1 from changing instantaneously and diode D_1 will not immediately turn off. So, both diodes D_1 and D_2 remain ON for an overlap period called the commutation time t_c.

With both diodes turned on, a closed circuit is established which involves both branches. The effective circuit voltage $V_C = (V_2 - V_1)$, called the **commutation voltage**, then drives a circulating current i_c, called the commutation current, through the two branches which have a total circuit inductance of $L_C = (L_1 + L_2)$.

In this idealized circuit, the volt drop across the diodes and the circuit resistance have been ignored. From basic electrical theory of inductive circuits, the current i_c increases with time at a rate dependent on the circuit inductance. The magnitude of the commutation current may be calculated from the following equations.

$$(V_2 - V_1) = (L_1 + L_2)\frac{di_c}{dt}$$

$$V_c = L_c\frac{di_c}{dt}$$

$$\frac{di_c}{dt} = \frac{V_c}{L_c}$$

If the commutation starts at a time t_1 and finishes at a time t_2, the magnitude of the commutation current I_C at any time t, during the commutation period, may be calculated by integrating the above equation from time t_1 to t.

$$I_c = \frac{1}{L_c}\int V_c\,dt$$

During the commutation period:

- It is assumed that the overall current through the circuit remains constant.

$$I = (I_1 + I_2)\ \ constant$$

As the circulating commutation current increases:

- Current (I_2) through the diode that is turning on **increases** in value

$$I_2 = I_c\ \ increasing$$

- Current (I_1) through the diode that is turning off **decreases** in value

$$I_1 = I - I_c \text{ decreasing}$$

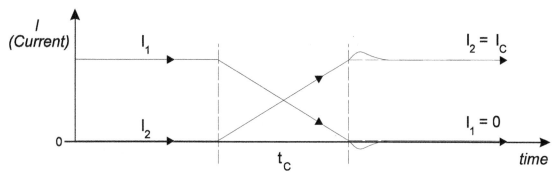

Figure 3.7:
The currents in each branch during commutation

For this special example, it can be assumed that the commutation voltage V_C is constant during the short period of the commutation. At time t the integration yields the following value of I_C, which increases linearly with time.

$$I_c = \frac{V_c}{L_c}(t - t_1)$$

When I_C has increased to a value equal to the load current I at time t_2, then all the current has been transferred from branch 1 to branch 2 and the current through the switch that is turning off has decreased to zero. The commutation is then over.

Consequently, at time $t2$

$$I_1 = 0$$

$$I_2 = I_c = I$$

At the end of commutation when $t = t2$, putting I_C equal to I in the above equation, the time taken to transfer the current from one circuit branch to the other (commutation time), may be calculated.

$$I = \frac{V_c(t_2 - t_1)}{L_c}$$

$$I = \frac{V_c t_c}{L_c}$$

$$t_c = \frac{I L_c}{V_c}$$

It is clear from this equation that the commutation time t_c depends on the overall circuit inductance ($L_1 + L_2$) and the commutation voltage.

From this we can conclude that:
- A large circuit inductance will result in a long commutation time.
- A large commutation voltage will result in a short commutation time.

In practice, a number of deviations from this idealized situation occur.
- The *diodes* are not ideal and do not turn off immediately when the forward voltage becomes negative. When a *diode* has been conducting and is then presented with a reverse voltage, some reverse current can still flow for a few microseconds as indicated in Figure 3.7. The current I_1 continues to decrease beyond zero to a negative value before returning to zero. This is due to the free charges that must be removed from the PN junction before blocking is achieved.
- Even if the commutation time is very short, the commutation voltage of an AC fed rectifier bridge does not remain constant but changes slightly during the commutation period. An increasing commutation voltage will tend to reduce the commutation time.

In practical power electronic converter circuits, commutation follows the same basic sequence outlined above. The figure below shows a typical 6-pulse rectifier bridge circuit to convert 3-phase AC currents I_A, I_B and I_C, to a DC current I_D.

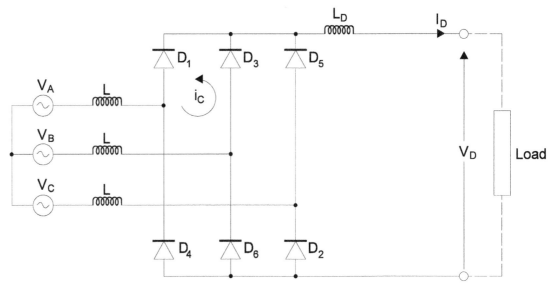

Figure 3.8:
3-Phase commutation with a 6-pulse diode bridge

This type of circuit is relatively simple to analyze because only 2 of the 6 diodes conduct current at any one time. The idealized commutation circuit can easily be identified. In this example, commutation is assumed to be taking place from diode D_1 to D_3 in the positive group, while D_2 conducts in the negative group.

In power electronic bridge circuits, it is conventional to number the diodes D_1 to D_6 in the sequence in which they turned ON and OFF. When V_A is the highest voltage and V_C the lowest, D_1 and D_2 are conducting.

In a similar way to the idealized circuit in Figure 3.6, when V_B rises to exceed V_A, D_3 turns on and commutation transfers the current from diode D_1 to D_3. As before, the commutation time is dependent on the circuit inductance (L) and the commutation voltage ($V_B - V_A$).

As can be seen from the 6-pulse diode rectifier bridge example above in Figure 3.8, commutation is usually initiated by external changes. In this case, commutation is controlled by the 3-phase supply line voltages. In other applications, commutation can also be initiated or controlled by other factors, depending on the type of converter and the application. Therefore, converters are often classified in accordance with the source of the external changes that initiate commutation.

- In the above example, the converter is said to be **line commutated** because the source of the commutation voltage is on the mains supply line.
- A converter is said to be **self-commutated** if the source of the commutation voltage comes from within the converter itself. Gate-commutated converters are typical examples of this.

3.6.1   Line commutated diode rectifier bridge

One of the most common circuits used in power electronics is the 3-phase line commutated 6-pulse rectifier bridge, which comprises 6 diodes in a *bridge connection*. Single-phase bridges will not be covered here because their operation can be deduced as a simplification of the 3-phase bridge.

In the analysis of the various types of converter that follow, the procedure will be to assume initially that the conditions and components are ideal. Once the principles have been established, any deviations from the ideal will be discussed. The following *ideal* assumptions are made:

- The supply voltages are '*stiff*' and completely sinusoidal
- Commutations are instantaneous and have no recovery problems
- Load currents are completely smooth
- Transformers and other line components are linear and ideal
- There is no volt drop in power electronic switches

These assumptions are made to gain an understanding of the circuits and to make estimates of currents, voltages, commutation times, etc. Thereafter, the limiting conditions that affect the performance of the practical converters and their deviation from the ideal conditions will be examined to bridge the gap from the ideal to the practical.

In the diode bridge, the diodes are not controlled from an external control circuit. Instead, commutation is initiated externally by the changes that take place in the supply line voltages, hence the name **line commutated rectifier**.

According to convention, the diodes are labeled D_1 to D_6 in the sequence in which they are turned ON and OFF. This sequence follows the sequence of the supply line voltages.

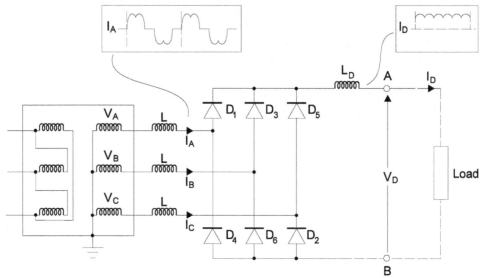

Figure 3.9:
Line commutated diode rectifier bridge

The 3-phase supply voltages comprise 3 sinusoidal voltage waveforms 120° apart which rise to their maximum value in the sequence A – B – C. According to convention, the phase-to-neutral voltages are labeled V_A, V_B and V_C and the phase-to-phase voltages are V_{AB}, V_{BC} and V_{CA}, etc.

These voltages are usually shown graphically as a vector diagram, which rotates counter-clockwise at a frequency of 50 times per second. A vector diagram of these voltages and their relative positions and magnitudes is shown below. The sinusoidal voltage waveforms of the supply voltage may be derived from the rotation of the vector diagram.

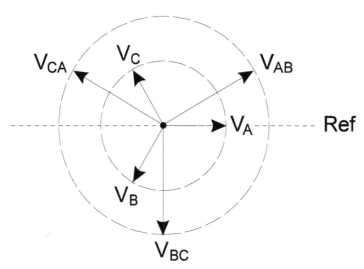

Figure 3.10:
Vector diagram of the 3-phase mains supply voltages

The output of the converter is the rectified DC voltage V_D, which drives a DC current I_D through a load on the DC side of the rectifier. In the idealized circuit, it is assumed that the DC current I_D is constant and completely smooth and without ripple.

The bridge comprises two commutation groups, one connected to the positive leg, consisting of diodes D_1–D_3–D_5, and one connected to the negative leg, consisting of diodes D_4–D_6–D_2. The commutation transfers the current from one diode to another in sequence and each diode conducts current for 120° of each cycle as shown in Figure 3.11.

In the upper group, the positive DC terminal follows the highest voltage in the sequence V_A–V_B–V_C via diodes D_1–D_3–D_5. When V_A is near its positive peak, diode D_1 conducts and the voltage of the +DC terminal follows V_A. The DC current flows through the load and returns via one of the lower group diodes. With the passage of time, V_A reaches its sinusoidal peak and starts to decline. At the same time, V_B is rising and eventually reaches a point when it becomes equal to and starts to exceed V_A. At this point, the forward voltage across diode D_3 becomes positive and it starts to turn on. The commutating voltage in this circuit, V_B–V_A starts to drive an increasing commutation current though the circuit inductances and the current through D_3 starts to increase as the current in D_1 decreases. In a sequence of events similar to that described above, commutation takes place and the current is transferred from diode D_1 to diode D_3. At the end of the commutation period, diode D_1 is blocking and the +DC terminal follows V_B until the next commutation takes place to transfer the current to diode D_5. After diode D_5, the commutation transfers the current back to D_1 and the cycle is repeated.

In the lower group, a very similar sequence of events takes place, but with negative voltages and the current flowing from the load back to the mains. Initially, D_2 is assumed to be conducting when V_C is more negative than V_A. As time progresses, V_A becomes equal to V_C and then becomes more negative. Commutation takes place and the current is transferred from diode D_2 to D_4. Diode D_2 turns off and D_4 turns on. The current is later transferred to diode D_6, then back to D_2 and the cycle is repeated.

In Figure 3.11, the conducting periods of the diodes in the upper and lower groups are shown over several cycles of the 3-Phase supply. This shows that only 2 diodes conduct current at any time (except during the commutation period, which is assumed to be infinitely short!!) and that each of the 6 diodes conducts for only one portion of the cycle in a regular sequence. The commutation takes place alternatively in the top group and the bottom group.

The DC output voltage V_D is not a smooth voltage and consists of portions of the *phase-to-phase voltage waveforms*. For every cycle of the 50 Hz AC waveform (20 msec), the DC voltage V_D comprises portions of the 6 voltage pulses, V_{AB}, V_{ac}, V_{BC}, V_{BA}, V_{CA}, V_{CB}, etc, hence the name 6-pulse rectifier bridge.

The average magnitude of the DC voltage may be calculated from the voltage waveform shown above. The average value is obtained by integrating the voltage over one of the repeating 120° portions of the DC voltage curve. This integration yields an average magnitude of the voltage V_D as follows.

$$V_D = 1.35 \times (RMS - Phase\ Voltage)$$

$$V_D = 1.35 \times V_{RMS}$$

For example, if $V_{RMS} = 415$ volts, $V_D = 560$ volts DC

If there is sufficient inductance in the DC circuit, then the DC current I_D will be fairly steady and the AC supply current will comprise segments of DC current from each diode in sequence. As an example, the current in the A-phase is shown in Figure 3.9. The non-sinusoidal current that flows in each phase of the supply mains can affect the performance of other AC equipment connected to the supply line that are designed to operate with sinusoidal waveforms. The effects of the non-sinusoidal currents is fully covered in Chapter 4: Electromagnetic compatibility (EMC).

In practice, to ensure that the diode reverse blocking voltage capability is properly specified, it is necessary to know the magnitude of the reverse blocking voltage which appears across each of the diodes. Theoretically, the maximum reverse voltage across a diode is equal to the peak of the phase–phase voltage. For example, the reverse voltage V_{CA} and V_{CB} appears across diode D_5 during the blocking period. In practice, a factor of safety of 2.5 is commonly used for specifying the reverse blocking capability of diodes and other power electronic switches. On a rectifier bridge fed from a 415 V power supply, the reverse blocking voltage V_{bb} of the diode must be higher than 2.5×440 V = 1100 V. Therefore, it is common practice to use diodes with a reverse blocking voltage of 1200 V.

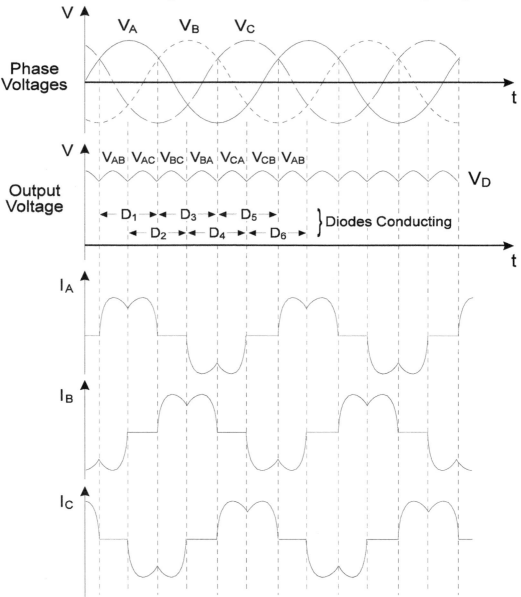

Figure 3.11
Voltage and current waveforms during commutation

3.6.2 The line commutated thyristor rectifier bridge

The output DC voltage and operating sequence of the diode rectifier above is dependent on the continuous changes in the supply line voltages and is not dependent on any control circuit. This type of converter is called an *uncontrolled diode rectifier bridge* because the DC voltage output is not controlled and is fixed at $1.35 \times V_{RMS}$.

If the diodes are replaced with thyristors, it then becomes possible to control the point at which the thyristors are triggered and therefore the magnitude of the DC output voltage can be controlled. This type of converter is called a *controlled thyristor rectifier bridge* and requires an additional control circuit to trigger the thyristor at the right instant. A typical 6-pulse thyristor converter is shown in Figure 3.12.

From the previous chapter, the conditions required before a thyristor will conduct current in a power electronic circuit are:

- A Forward Voltage must exist across the thyristor

AND

- A Positive Pulse must be applied to the thyristor gate

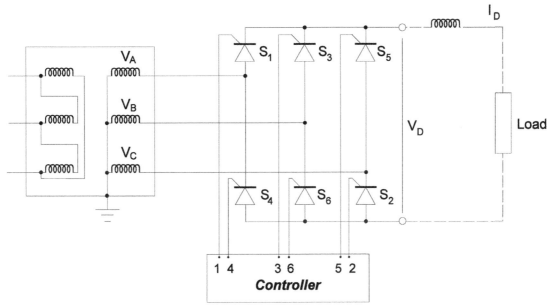

Figure 3.12:
6-pulse controlled thyristor rectifier bridge

If each thyristor were triggered at the instant when the forward voltage across it tends to become positive, then the thyristor rectifier operates in the same way as the diode rectifier described above. All the voltage and current waveforms of the diode bridge apply to the thyristor bridge. A thyristor bridge operating in this mode is said to be operating with a *zero delay angle* and gives a voltage output of:

$$V_D = 1.35 \times V_{RMS}$$

The output of the rectifier bridge can be controlled by delaying the instant at which the thyristor receives a triggering pulse. This delay is usually measured in degrees from the point at which the switch CAN turn on, due to the forward voltage becoming positive.

The angle of delay is called the *delay angle*, or sometimes the *firing angle*, and is designated by the symbol α. The reference point, for the angle of delay, is the point where a phase voltage wave crosses the voltage of the previous phase and becomes positive relative to it. A diode rectifier can be thought of as a converter with a *delay angle of α = 0°*.

The main purpose of controlling a converter is to control the magnitude of the DC output voltage. In general, the bigger the delay angle, the lower the average magnitude of the DC voltage. Under steady state operation of a controlled thyristor converter, the delay angle for each switch is the same. Figure 3.13 shows the voltage waveforms where the triggering of the switches has been delayed by an angle of α degrees.

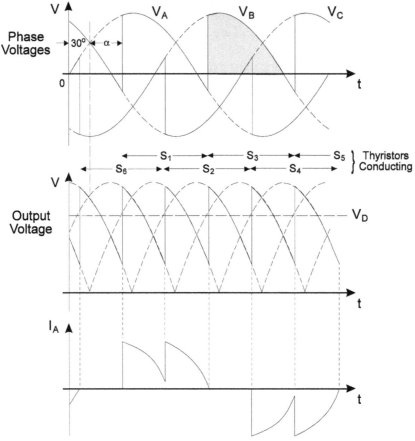

Figure 3.13:
Voltage waveforms of a controlled rectifier

In the positive switch group, the positive DC terminal follows the voltage associated with the switch that is in conduction in the sequence V_A–V_B–V_C. Assume, initially, that thyristor S_1 associated with voltage V_A is conducting and S_3 is not yet triggered. The voltage on the + **bus** on the DC side follows the declining voltage V_A because, in the absence of S_3 conduction, there is still a forward voltage across S_1 and it will continue to conduct. When S_3 is triggered after a delay angle = α, the voltage on + **bus** jumps to V_B, whose value it then starts to follow. At this instant, with both S_1 and S_3 conducting, a negative commutation voltage equal to V_B–V_A appears across the switch S_1 for the commutation period, which then starts to turn off. With the passage of time, V_B reaches its sinusoidal peak and starts to decline, followed by + DC terminal. At the same time, V_C is rising and when S_5 is triggered, the same sequence of events is repeated and the current is commutated to S_5.

As with the diode rectifier, the average magnitude of the DC voltage V_D can be calculated by integrating the voltage waveform over a $120°$ period representing a repeating portion of the DC voltage. At a delay angle α, the DC voltage is given by:

$$V_D = 1.35 \times (RMS - Phase\ Voltage) \times Cos\ \alpha$$

$$V_D = 1.35 \times V_{RMS} \times Cos\ \alpha$$

This formula shows that the theoretical DC voltage output of the thyristor rectifier with a firing angle $\alpha = 0$ is the same as that for a diode rectifier. It also shows that the average value of the DC voltage will decrease as the delay angle is increased and is dependent on the cosine of the delay angle. When $\alpha = 90°$, then $Cos\alpha = 0$ and $V_D = 0$, which means that the average value of the DC voltage is zero. The instantaneous value of the DC voltage is a *saw-tooth* voltage as shown in the figure below.

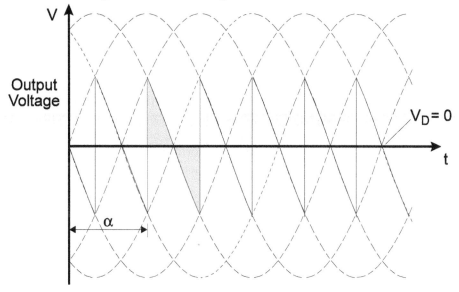

Figure 3.14:
DC output voltage for delay angle $\alpha = 90°$

If the delay angle is increased further, the average value of the DC voltage becomes negative. In this mode of operation, the converter operates as an *inverter*. It is interesting to note that the direction of the DC current remains unchanged because the current can only flow through the switches in the one direction. However, with a negative DC voltage, the direction of the power flow is reversed and the power flows from the DC side to the AC side. Steady state operation in this mode is only possible if there is a voltage source on the DC side. The instantaneous value of the DC voltage for $\alpha > 90°$ is shown in Figure 3.15.

In practice, the commutation is not instantaneous and lasts for a period dependent on the circuit inductance and the magnitude of the commutation voltage. As in the idealized case, it is possible to estimate the commutation time from the commutation circuit inductance and an estimate of the average commutation voltage.

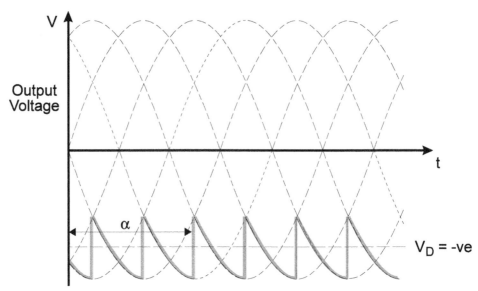

Figure 3.15:
DC voltage when the delay angle α > 90°

As in the diode rectifier, the steady DC current I_D comprises segments of current from each of the 3 phases on the AC side. On the AC side, the current in each phase comprises non-sinusoidal blocks, similar to those associated with the diode rectifier and with similar harmonic consequences. In the case of the diode bridge, with a delay angle of α = 0, the angle between the phase current and the corresponding phase voltage on the AC side is roughly zero. Consequently, the power factor is roughly unity and converter behaves something like a resistive load.

For the controlled rectifier, with a delay angle of α, the angle between the phase current and the corresponding phase voltage is also roughly α, but normally called the power factor angle ∅. This angle should be called the *displacement factor* because it does not really represent power factor (see later). Consequently, when the delay angle of the thyristor rectifier is changed to reduce the DC voltage, the angle between the phase current and voltage also changes by the same amount. The converter then behaves like a resistive-inductive load with a displacement factor of Cos∅. It is well known that the *power factor* associated with a controlled rectifier falls when the DC output voltage is reduced.

Delay angle	Converter behavior
α = 0°	Behaves like a resistive load
0° < α < 90°	Behaves like a resistive/inductive load and absorbs active power
α = 90°	Behaves like an inductive load with no active power drawn
α > 90°	Behaves like an inductive load but is also a source of active power

A common example of this is a DC motor drive controlled from a thyristor converter. As the DC voltage is reduced to reduce the DC motor speed at constant torque, the power factor drops and more reactive power is required at the supply line to the converter.

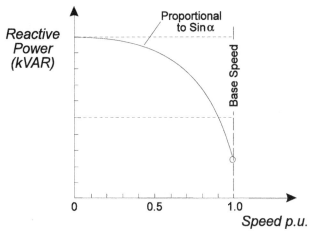

Figure 3.16:
Reactive power requirements of a DC motor drive with a constant torque load fed from a line-commutated converter

Figure 3.17 summarizes the possible vector relationships between the phase voltage and the fundamental component of the phase current in the supply line for the various values of delay angle α.

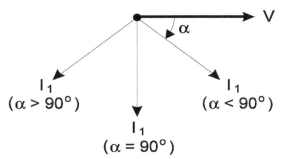

Figure 3.17:
Vector diagram of phase voltage and fundamental current for a controlled thyristor rectifier bridge

The phase current on the AC side is, fundamentally, a non-sinusoidal *square wave*. By applying the principles of harmonic analysis, using the Fourier transform, this non-sinusoidal wave can be resolved into a fundamental (50 Hz) sinusoidal wave plus a number of sinusoidal harmonics (refer to Chapter 4). The fundamental waveform has the highest amplitude and therefore the most influence on the power supply system. In a 6-pulse rectifier bridge, the 5th harmonic has the highest magnitude, theoretically 20% of the fundamental current.

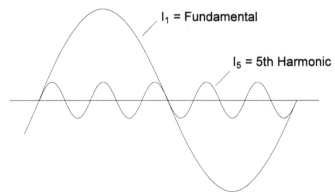

Figure 3.18:
The fundamental current and the 5th harmonic current

The RMS value of the fundamental current can be calculated from the following formula, which is derived from fundamental principles:

$$I_1 = \sqrt{3}\,\frac{\sqrt{2}}{\pi}\,I_D = 0.78\,I_D \quad \text{amps}$$

The corresponding apparent power S_1 kVA is given by:

$$S_1 = \sqrt{3}\ V_{RMS}\,I_1 \ \ \text{kVA}$$

$$S_1 = \sqrt{3}\ V_{RMS}\ 0.78\,I_D \ \ \text{kVA}$$

$$S_1 = 1.35\ V_{RMS}\,I_D \ \ \text{kVA}$$

The active power component is given by:

$$P_1 = S_1\,\text{Cos}\,\varphi \ \ \text{kW}$$

$$P_1 = 1.35\ V_{RMS}\,I_D \ \ \text{kW}$$

This confirms that the active power calculated on the AC side is identical to the power calculated for the DC side ($V_D.I_D$), since from the previous formula $V_D = 1.35\ V_{RMS}\text{Cos}\alpha$.
The reactive power component is given by:

$$Q_1 = S_1\,\text{Sin}\,\varphi \ \ \text{kVAr}$$

$$Q_1 = 1.35\ V_{RMS}\,I_D\,\text{Sin}\,\varphi \ \ \text{kVAr}$$

This formula illustrates that, if the load current is held constant (constant torque load on a DC motor), the reactive power will increase in proportion to Sinα as the triggering *delay angle* is increased.

Looking at the rectifier from the 3-phase supply, an effective phase-to-phase short circuit occurs across the associated supply lines during commutation, when the 2 sequential switches are conducting. For example, when switch S_3 is triggered and switch S_1 continues to conduct, the voltage of V_A and V_B must be equal at switches themselves (except for the small volt drop across the switches). The commutation voltage V_B–V_A

drives a circulating current through S_1 and S_3 and the circuit inductance 2L. Depending on the *delay angle*, the commutation voltage can be quite large. At the voltage source, the magnitude of the voltages V_A and V_B are depressed during this period by an amount dependent on the circulating current and circuit inductance. This additional non-desirable effect in the supply line is called *voltage notching*. The effect of notching is to slightly reduce the DC voltage V_D, but this reduction is very small and may be ignored. However, notching is important when considering the losses in the converter.

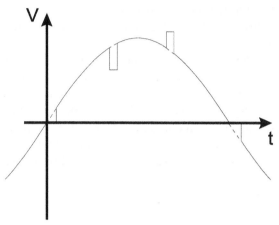

Figure 3.19:
Voltage notching in the supply line

3.6.3   Practical limitations of line commutated converters

The above analysis covers the theoretical aspects of uncontrolled and controlled converters. In practice, the components are not ideal and the commutations are not instantaneous. This results in certain deviations from the theoretical performance.

One of the most important deviations is that the DC load current is never completely smooth. The reason for this is fairly obvious. Accepting that the instantaneous DC voltage V_D can never be completely smooth, if the load is purely *resistive*, the DC load current cannot be completely smooth because it will linearly follow the DC voltage. Also, at delay angles $\alpha > 60°$, the DC output voltage becomes discontinuous and, consequently, so would the DC current. In an effort to maintain a smooth DC current, practical converters usually have some inductance L_D in series with the load on the DC side. For complete smoothing, the value of L_D should theoretically be infinite, which is not really practical.

The practical consequence of this is that the theoretical formula for the calculated value of DC voltage ($V_D = 1.35\ V_{RMS}\ \mathrm{Cos}\alpha$) is not completely true for all values of delay angle α. Practical measurements confirm that it only hold true for delay angles up to about 75°, but this depends on the type of load and, in particular, the DC load inductance. Experience shows that for a particular delay angle $\alpha > 60°$, the average DC voltage will be higher than the theoretical value as shown in the figure below.

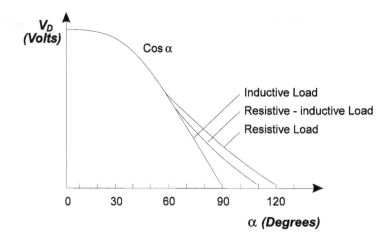

Figure 3.20:
Deviation of DC voltage from theoretical vs delay angle

3.6.4 Applications for line commutated rectifiers

An important application of the line-commutated converter is the DC motor drive. The figure below shows a single *controlled* line-commutated converter connected to the armature of a DC Motor. The converter provides a variable DC voltage V_A to the armature of the motor, controlled from the control circuit of the converter.

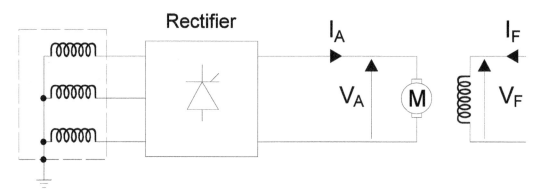

Figure 3.21:
Converter fed DC motor drive

When the delay angle is less than 90°, the DC voltage is positive and a positive current I_A flows into the armature of the DC motor to deliver active power to the load. The drive system is said to be operating in the *1st quadrant* where the motor is running in the forward direction with active power being transferred from the supply to the motor and its mechanical load.

The motor field winding is usually separately excited from a simple diode rectifier and carries a field magnetizing current I_F. For a fixed field current, the speed of the motor is proportional to the DC voltage at the armature. The speed can be controlled by varying the delay angle of the converter and its output armature voltage V_A.

If the delay angle of the converter is increased to an angle greater than 90°, the voltage V_D will become negative and the motor will slow to a standstill. The current I_D also reduces to zero and the supply line can be disconnected from the motor without breaking any current.

Consequently, to stop a DC motor, the delay angle must be increased to value sufficiently larger than 90° to ensure that the voltage V_D becomes negative. With V_D negative and I_D still positive, the converter transiently behaves like an inverter and transfers active power from the motor to the supply line. This also acts as a brake to slow the motor and its load quickly to standstill. In this situation, the drive system is said to be operating in the *2nd quadrant* where the motor is running in the forward direction but the active power is being transferred back from the motor to the supply line.

The concept of the 4 operating quadrants has been covered in Chapter 1, but is illustrated again below. It illustrates the 4 possible operating states of any drive system and also shows the directions of V_D and I_D for the DC motor drive application.

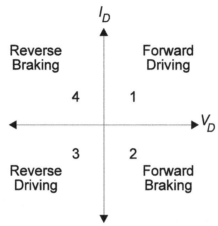

Figure 3.22:
Operating quadrants for variable speed drives

The converters discussed so far have been *single converters*, which are only able to operate with positive DC current (I_D = +ve), which means that the motor can only run in the forward direction but active power can be transferred in either direction. Single DC converters can only operate in quadrants 1 & 4 and are known as 2 Quadrant converters.

To operate in quadrants 3 & 2, it must be possible to reverse the direction of I_D. This requires an additional converter bridge connected for current to flow in the opposite direction. This type of converter is known as a *4 quadrant DC converter*, and sometimes also called a double or back-to-back 6-pulse rectifier.

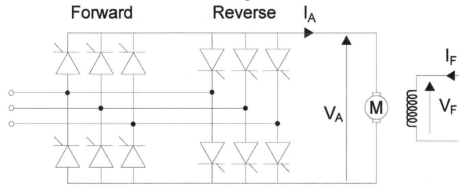

Figure 3.23:
4-quadrant line-commutated rectifier

With a DC motor drive fed from a 4-quadrant DC converter, operation in all 4 quadrants is possible with speed control in either the forward or reverse direction. A change of direction of the motor can quickly be achieved. Converter-1 is used as a controlled rectifier for speed control in the forward direction of rotation, while converter-2 is blocked, and vice versa in the reverse direction.

Assume, initially, that the motor is running in the forward direction under the control of Converter-1 with a *delay angle* of < 90°. Converter-2 is blocked. The changeover sequence from running in the forward direction to the reverse direction is as follows:

- Converter-1 delay angle increased to $\alpha > 90°$. This means that DC voltage V_D < 0 and DC current I_D is decreasing.
- When $I_D = 0$, Converter-1 is blocked and thyristor firing is terminated.
- After small delay, converter-2 unblocked and starts in the inverter mode with a firing angle greater than 90°.
- If the motor is still turning in the forward direction, converter-2 DC current I_D starts to increase in the negative direction and the DC machine acts as a generator and is braked to standstill, returning energy to the supply line.
- As the firing angle is reduced $\alpha < 90°$, converter-2 changes from the inverter to rectifier mode and, as voltage V_D increases, the motor starts to rotate in the opposite direction.

In a DC motor drive, reversal of the direction of rotation can also be achieved by using a single converter and changing the direction of the excitation current. This method can only be used where there are no special drive requirements for changing over from forward to reverse operation. In this case, the changeover is done mechanically using switches in the field circuit during a period at standstill. Considerable time delays are required during standstill to remagnetize the field in the reverse direction.

There are many practical applications for both uncontrolled and controlled line-commutated rectifiers. Some of the more common applications include the following:

- DC motor drives with variable speed control
- DC supply for variable voltage variable frequency inverters
- Slip-energy recovery converters for wound rotor induction motors
- DC excitation supply for machines
- High voltage DC converters
- Electrochemical processes

3.7 Gate commutated inverters (DC/AC converters)

Most modern AC variable speed drives in the 1 kW to 500 kW range are based on *gate-commutated devices* such as the GTO, MOSFET, BJT and IGBT, which can be turned ON and OFF by low power control circuits connected to their control gates.

The difficulties experienced with thyristor commutation in the early days of PWM inverters have largely been overcome by new developments in power electronic technology. Diodes and thyristors are still used extensively in line-commutated rectifiers.

Starting with a DC supply and using these semiconductor *power electronic switches*, it is not possible to obtain a pure sinusoidal voltage at the load. On the other hand, it may be possible to generate a near-sinusoidal current. Consequently, the objective is to control these switches in such a way that the current through the inductive circuit should approximate a sinusoidal current as closely as possible.

3.7.1   Single-phase square wave inverter

To establish the principles of gate-controlled inverter circuits, the figure below shows four semiconductor power switches feeding an inductive load from a single-phase supply.

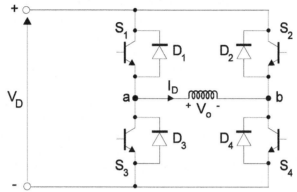

Figure 3.24:
Single-phase DC to AC inverter

This circuit can be considered to be an electronic reversing switch, which allows the input DC voltage V_D to be connected to the inductive load in any one of the following ways:

(1) S_1 = on, S_4 = on, giving $+V_D$ at the load
(2) S_2 = on, S_3 = on, giving $-V_D$ at the load
(3) S_1 = on, S_2 = on, giving zero volts at the load
    S_3 = on, S_4 = on, giving zero volts at the load
(4) S_1 = on, S_3 = on, giving a short circuit fault
    S_2 = on, S_4 = on, giving a short circuit fault

However, these four switches can be controlled to give a square waveform across the inductive load as shown in Figure 3.24. This makes use of switch configuration (1) and (2), but not switch configuration (3) or (4). Clearly, for continued safe operation, option (4) should always be avoided. In the case of a purely inductive load, the current waveform is a triangular waveform as shown in the Figure 3.25.

In the first part of the cycle, the current is negative although only switches S_1 and S_4 are on. Since most power electronic devices cannot conduct negatively, to avoid damage to the switches, this negative current would have to be diverted around them. Consequently, diodes are usually provided in anti-parallel with the switches to allow the current flow to continue. These diodes are sometimes called *reactive* or *free-wheeling diodes* and conduct whenever the voltage and current polarities are opposite. This occurs whenever there is a reverse power flow back to the DC supply.

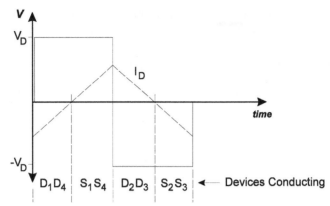

Figure 3.25:
Square wave modulation waveforms

The frequency of the periodic square wave output is called the fundamental frequency. Using Fourier analysis, any repetitive waveform can be resolved into a number of sinusoidal waveforms, comprising one sinusoid at fundamental frequency plus a number of sinusoidal harmonics at higher frequencies, which are multiples of the fundamental frequency. The harmonic spectrum for a single-phase square wave output is shown in the figure below. The amplitude of the higher order harmonics voltages falls off rapidly with increasing frequency.

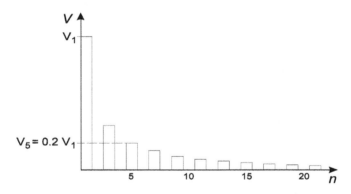

Figure 3.26:
Square-wave harmonic spectrum

The RMS value of the fundamental sinusoidal voltage component is:

$$V_1 = 2\frac{\sqrt{2}}{\pi}V_d \quad \text{volts}$$

The RMS value of the nth harmonic voltage:

$$V_n = \frac{V_1}{n} \quad \text{volts}$$

This illustrates that the square wave output voltage has a lot of unwanted components of reasonably large magnitude at frequencies close to the fundamental. The current flowing in the load as a result of the output voltage is distorted, as demonstrated by the non-sinusoidal current wave-shape. In this example, the current has a triangular shape.

If the square-wave voltage were presented to a single-phase induction motor, the motor would run at the frequency of the square-wave but, being a linear device (inductive/resistive load), it would draw non-sinusoidal currents and would suffer additional heating due to the harmonic currents. These currents may also produce pulsating torques.

To change the speed of the motor, the fundamental frequency of the inverter output can be changed by adjusting the speed of the switching. To increase frequency, switching speed can be increased and to decrease frequency, switching speed can be decreased.

If it is required to also control the magnitude of the output voltage, the average inverter output voltage can be reduced by inserting periods of zero voltage, using switch configuration (3) as shown in Figure 3.24. Each half cycle then consists of a square pulse which is only a portion of a half period as shown in the figure below.

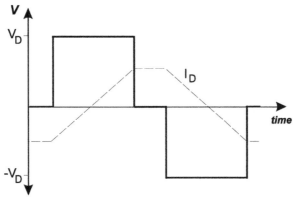

Figure 3.27:
Square wave modulation with reduced voltage pulse width

The process of changing the width of the pulse to reduce the average RMS value of a waveform is called pulse width modulation (PWM). In the single phase example in Figure 3.27 above, pulse width modulation makes it possible to control the RMS value of the output voltage. The fundamental sinusoidal component of voltage is continuously variable in the following range:

$$zero \quad - \quad 2\frac{\sqrt{2}}{\pi}V_D \quad volts$$

The harmonic spectrum of this modified waveform depends on the fraction that the pulse is of the full square wave, but is broadly similar to the waveform shown in Figure 3.26.

3.7.2   Single-phase pulse width modulation (PWM) inverter

The fact that the voltage supply to the stator of an AC induction motor is a square wave and is distorted is not in itself a problem for the motor. The main problem comes from the distortion of the current waveform, which results in extra copper losses and shaft torque pulsations. The ideal inverter output is one that results in a current waveform of low harmonic distortion.

Since an AC induction motor is predominantly inductive, with a reactance that depends on the frequency ($X_L = j2\pi fL$), it is beneficial if the **voltage harmonic distortion** can be pushed into the high frequencies, where the motor impedance is high and not much distorted current will flow.

One technique for achieving this is *sine-coded pulse width modulation (sine-PWM)*. This requires the power devices to be switched at frequencies much greater than that of the fundamental frequency producing a number of pulses for each period of the desired output period. The frequency of the pulses is called the *modulation frequency*. The width of the pulses is varied throughout the cycle in a sinusoidal manner giving a voltage waveform as shown in Figure 3.28. This figure also shows the current waveform for an inductive load showing the improvement in the waveform.

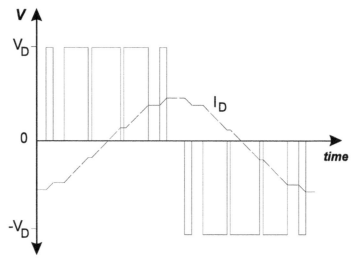

Figure 3.28:
Sine-coded pulse width modulated voltage and current

The improvement in the current waveform can be explained by the harmonic spectrum shown in Figure 3.29. It can be seen that, although the voltage waveform still has many distortion components, they now occur at higher harmonic frequencies, where the high load impedance of the motor is effective in reducing these currents.

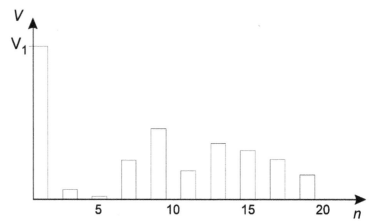

Figure 3.29:
Harmonic spectrum for a PWM inverter

Increasing the *modulation frequency* will improve the current waveform, but at the expense of increased losses in the switching devices of the inverter. The choice of modulation frequency depends on the type of switching device and its frequency. With the force-commutated thyristor inverter (10 years ago), a modulation frequency of up to 1 kHz was possible. With the introduction of GTOs and BJTs, this could be pushed up to around 5 kHz. With IGBTs, the modulation frequency could be as high as 20 kHz. In practice, a maximum modulation frequency of up to 12 kHz is common with IGBT inverters up to about the 22 kW motor size and 8 kHz for motors up to about 500 kW. The choice of modulation frequency is a trade off between the losses in the motor and in the inverter. At low modulation frequencies, the losses in the inverter are low and those in the motor are high. At high modulation frequencies, the losses in the inverter increase, while those in the motor decrease.

One of the most common techniques for achieving sine-coded PWM in practical inverters is the sine-triangle intersection method shown in Figure 3.30.

A triangular *saw-tooth* waveform is produced in the control circuit at the desired inverter switching frequency. This is compared in a comparator with a sinusoidal reference signal, which is equal in frequency and proportional in magnitude to that of the desired sinusoidal output voltage. The voltage V_{AN} (Figure 3.30(b)) is switched high whenever the reference waveform is greater than the triangle waveform. The voltage V_{BN} (Figure 3.30(c)) is controlled by the same triangle waveform but with a reference waveform shifted by 180°.

The actual phase-to-phase output voltage is then V_{AB} (Figure 3.30(d)), which is the difference between V_{AN} and V_{BN}, which consists of a series of pulses each of whose width is related to the value of the reference sine-wave at that time. The number of pulses in the output voltage V_{AB} is double that in the inverter leg voltage V_{AN}. For example, an inverter switching at 5 kHz should produce switching distortion at 10 kHz in the output phase-to-phase voltage. The polarity of the voltage is alternatively positive and negative at the desired output frequency.

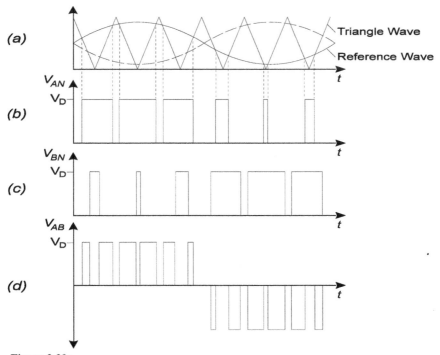

Figure 3.30:
Principle of triangle intersection PWM

It can also be seen that the reference sine-wave in Figure 3.30 is given a DC component so that the pulse produced by this technique has a positive width. This puts a DC bias on the voltage of each leg as shown in Figures 3.30 (b) & (c). However, each leg has the same DC offset which disappears from the load voltage.

The technique using sine-triangle intersection is particularly suited for use with the older analogue control circuits, where the two reference waveforms were fed into a comparator and the output of the comparator was used to trigger the inverter switches.

Modern *digital techniques* operate on the basis of a switching algorithm, for example by producing triggering pulses proportional to the area under a part of the sine wave. In recent times, manufacturers have developed a number of different algorithms that optimize the performance of the output waveforms for AC induction motors. These techniques result in PWM output waveforms which are similar to those shown in Figure 3.30.

The sine-coded PWM voltage waveform is a composite of a high frequency square wave at the pulse frequency (the switching carrier) and the sinusoidal variation of its width (the modulating waveform). It has been found that, for lowest harmonic distortion, the modulating waveform should be synchronised with the carrier frequency, so that is it should contain an integral number of carrier periods. This requirement becomes less important with high carrier frequencies of more than about twenty times the modulating frequency.

The voltage and frequency of a sinusoidal PWM waveform are varied by changing the reference waveform of Figure 3.30(a) giving outputs as shown in Figure 3.31.

- Figure 3.31(a) shows a base case, with the rated V/f ratio
- Figure 3.31(b) shows the case where the voltage reference is halved, resulting in the halving of each pulse
- Figure 3.31(c) shows the case where the reference frequency is halved, resulting in the extension of the modulation over twice as many pulses

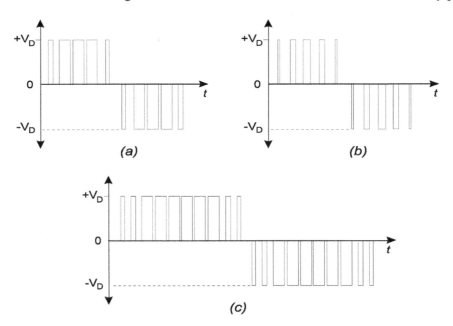

Figure 3.31:
Variation of frequency and voltage with sinusoidal PWM

The largest voltage with sine-coded PWM occurs when the pulses in the middle are widest, giving an output with a peak voltage equal to the supply. The modulation index is defined as the ratio of the peak AC output to the DC supply. Thus the largest output voltage occurs when the modulation index is 1. It is possible to achieve larger voltages than the DC supply by abandoning strict sine-PWM by adding some distortion to the sinusoidal reference voltage. This results in the removal of some of the pulses near the centre of the positive and negative parts of the waveform, a process called pulse dropping. In the limit, a square wave voltage waveform can be achieved with a peak value which is up to 127% of what can be achieved by strict sine-PWM.

3.7.3   Three-phase inverter

A three-phase inverter could be constructed from three inverters of the type shown in Figure 3.24. However, it is more economical to use a 6-pulse (three-leg) bridge inverter as shown in Figure 3.32.

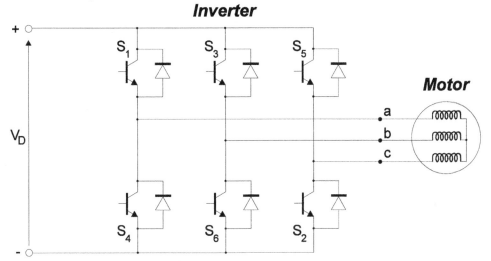

Figure 3.32:
Three-phase inverter using gate controlled switches

In its simplest form, a square output voltage waveform can be obtained by switching each leg high for one half-period and low for the next half-period, at the same time ensuring that each phase is shifted one third of a period (120°) as shown in the Figure 3.33. The resulting phase-to-phase voltage waveform comprises a series of square pulses whose widths are two thirds of the period of the switch in each phase. The resulting voltage waveform is called a *quasi-square wave (QSW)* voltage. This simple technique was used in early voltage source inverters (VSI) which used forced commutated thyristors in the inverter bridge. To maintain a constant V/f ratio, the magnitude of the DC bus voltage was controlled by the rectifier bridge to keep a fixed ratio to the output frequency, which was controlled by the inverter bridge. This technique was sometimes also called *pulse amplitude modulation (PAM)*.

The output voltage of a three-phase converter has a harmonic spectrum very similar to the single-phase square wave, except that the triplen harmonics (harmonics whose frequency is a multiple of three times the fundamental frequency) have been eliminated. In an inverter with a 3-phase output, this means that the 3rd, 9th, 15th, 21st, etc harmonics are eliminated. To develop a 3-phase variable voltage AC output of a particular frequency, the voltages V_{AN}, V_{BN}, V_{CN} on the 3 output terminals a, b, & c in Figure 3.32 can be modulated on and off to control both the voltage and the frequency. The pulse-width ratio over the period can be changed according to a sine-coded PWM algorithm.

When the phase-phase voltage V_{AB} is formed, the present modulation strategy gives only positive pulses for a half period followed by negative pulses for a half period, a condition known as *consistent pulse polarity*. It can be shown that *consistent pulse polarity* guarantees lowest harmonic distortion with most of the distortion being at twice the inverter chopping frequency. The presence of both positive and negative pulses throughout the whole period of the phase–phase voltage (*inconsistent pulse polarity*) gives distortion at the inverter chopping frequency, where it will have more effect on current distortion and is a sign of a poor modulation scheme.

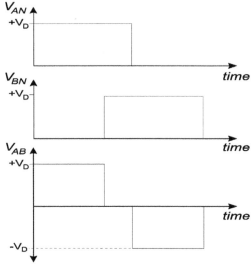

Figure 3.33:
Quasi square wave modulation output waveforms

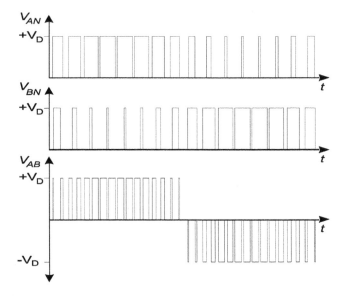

Figure 3.34:
Output voltage waveform of a 3-phase sine coded PWM

Manufacturers of AC frequency converters continue to work on the development of more efficient PWM algorithms in an attempt to improve the current waveform. The ultimate objective is a completely sinusoidal current, which produces no harmonic losses in the motor. These more advanced PWM algorithms have become possible as a result of the increased speed and power of microprocessors. Most reputable PWM inverters can operate at modulation frequencies between 2 kHz and 16 kHz and produce a current waveform, which is sufficiently sinusoidal to overcome the problem of motor de-rating for harmonic losses. However, as a result of the high PWM frequencies, a new problem has emerged, the high frequency leakage current due to the motor cable capacitance. This issue is covered in Chapter 4.

In practical inverters, there are two conflicting requirements which need to be met when it is required to accelerate a motor from standstill to rated speed with constant V/f ratio.

- The need to operate the inverter at its highest possible switching frequency to achieve low current distortion
- The importance of maintaining synchronization

A common strategy to achieve both, particularly for older PWM inverters, is to begin with the inverter switching frequency at about half the maximum value. As the speed is increased, the saw-tooth carrier frequency is increased in proportion to maintain synchronism. When the carrier frequency reaches its maximum, it is then switched to half its value for further increase in the output frequency.

Thus the inverter exhibits a continual ramp increase in frequency followed by a sudden reduction at the changeover point. If the inverter is operating in the audible range then a change in pitch will be heard similar to the sound of a car engine as the car accelerates through the gears, hence the term '*gear-changing*'.

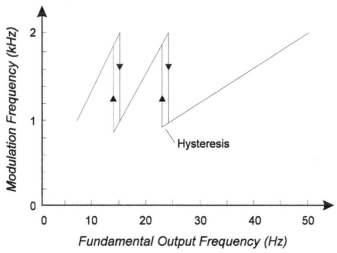

Figure 3.35:
Changing the modulation frequency in steps (gear-changing)

When the motor speed is reduced from maximum to zero, there is a similar change in carrier frequency with output frequency. However, the changeover points must be different, otherwise an inverter sitting at one of the changeover frequencies might continually oscillate between the upper and lower carrier frequency. This is avoided by introducing hysteresis in the control scheme as shown in Figure 3.35.

3.8   Gate controlled power electronic devices

A number of gate controlled devices have become available in the past decade, which are suitable for use as bi-stable switches on power inverters for AC Variable Speed Drives. These can be divided into two main groups of components:

- Those based on thyristor technology such as gate turnoff thyristor (GTO) and field controlled thyristor (FCT)
- Those based on transistor technology such as the bipolar junction transistor (BJT), field effect transistor (FET) and the insulated gate bipolar transistor (IGBT)

3.8.1   Gate turn-off thyristor (GTO)

A *GTO thyristor* is another member of the thyristor family and is very similar in appearance and performance to a *normal thyristor*, with the important additional feature that it can be turned off by applying a negative current pulse to the gate. *GTO thyristors* have high current and voltage capability and are commonly used for larger converters, especially when self commutation is required.

SYMBOL:

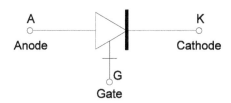

IDEAL: Forward conduction: Resistance (less)
Forward blocking: Loss (less) (no leakage current)
Reverse blocking: Loss (less) (no leakage current)
Switch on/off time: Instantaneous

The performance of a GTO is similar to a normal thyristor. Forward conduction is blocked until a positive pulse is applied to the gate terminal. When the GTO has been turned on, it behaves like a thyristor and continues to conduct even after the gate pulse is removed, provided that the current is higher than the holding current. The GTO has a higher forward voltage drop of typically 3 V to 5 V. Latching and holding currents are also slightly higher.

The important difference is that the GTO may be turned off by a negative current pulse applied to the gate terminal. This important feature permits the GTO to be used in self commutated inverter circuits. The magnitude of the off pulse is large and depends on the magnitude of the current in the power circuit. Typically, the gate current must be 20% of the anode current. Consequently, the triggering circuit must be quite large and this results in additional commutation losses. Like a thyristor, conduction is blocked in the reverse biased direction or if the holding current falls below a certain level.

Since the GTO is a special type of thyristor, most of the other characteristics of a thyristor covered above also apply to the GTO and will not be repeated here. The mechanical construction of a GTO is very similar to a normal thyristor with stud types common for smaller units and disc types common for larger units.

GTO thyristors are usually used for high voltage and current applications and are more robust and tolerant to over-current and over-voltages than power transistors. GTOs are available for ratings up to 2500 amps and 4500 volts. The main disadvantages are the high gate current required to turn the GTO off and the high forward volt drop.

Power electronic converters of all types are usually controlled by an electronic control circuit which controls the on/off state of the power electronic devices and provides the interface for the external controls. Until recently, all control circuits were of the analog type using operational amplifiers (Op-Amps). Modern control circuits are usually of the **digital** type using microprocessors.

3.8.2 Field controlled thyristors (FCT)

Although the GTO is likely to maintain its dominance for the high power, self commutated converter applications for some time, new types of thyristor are under development in which the gate is voltage controlled. **Turn on** is controlled by applying a **positive voltage** signal to the gate and **turn off** by a **negative voltage**. Such a device is called a *field controlled thyristor (FCT)* and the name highlights the similarity to the field effect transistor (FET). The FCT is expected to eventually supersede the GTO because it has a much simpler control circuit in which both the cost and the losses may be substantially reduced. Small FCTs have become available and it is expected that larger devices will come into use in the next few years. Development of a practical cost effective device has been a bit slower than expected.

3.8.3 Power bipolar junction transistors (BJT)

Transistors have traditionally been used as amplification devices, where control of the base current is used to make the transistor conductive to a greater or lesser degree. Until recently, they were not widely used for power electronic applications. The main reasons were that the control and protective circuits were considerably more complicated and expensive and transistors were not available for high power applications. They also lacked the overload capacity of a thyristor and it is not feasible to protect transistors with fuses.

In the mid-1980s, the NPN transistor known as a *bipolar junction transistor (BJT)* has become a cost effective device for use in power electronic converters. Modern BJTs are usually supplied in an encapsulated module and each BJT has two **power terminals**, called the *collector (C)* and *emitter (E)*, and a third **control terminal** called the *base (B)*.

SYMBOL:

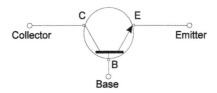

IDEAL: Forward conduction: Resistance (less)
               Forward blocking: Loss (less) (no leakage current)
               Reverse blocking: Loss (less) (no leakage current)
               Switch on/off time: Instantaneous

A transistor is not inherently a *bi-stable (on/off) device*. To make a transistor suitable for the conditions in a power electronic circuit where it is required to switch from the **blocking state (high voltage, low current)** to the **conducting state (low voltage, high current)** it must be used in its extreme conditions, fully off to fully on. This potentially stresses the transistor and the trigger and protective circuits must be co-ordinated to ensure the transistor is not permitted to operate outside its *safe operating area*.

Suitable control and protective circuits have been developed to protect the transistor against over-current when it is turned on and against over-voltage when it is turned off.

When turned on, the control circuit must ensure that the transistor does not come out of saturation otherwise it will be required to dissipate high power. In practice, the control system has proved to be cost effective, efficient, and reliable in service.

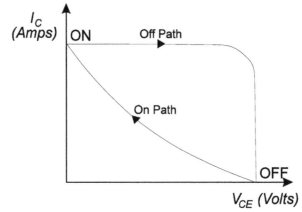

Figure 3.36:
Switching locus of a power BJT with an inductive load

Transistors do not tolerate reverse voltages. When BJTs are used in inverter bridges, they must be protected against high reverse voltages by means of a reverse diode in series or in parallel. For the same reason, transistors are not used in rectifier bridges, which have to be able to withstand reverse voltages.

In general, transistors were considered to be less robust and less tolerant of overloads and 'spikes' than thyristors. *GTO thyristors* were often preferred for converters. In spite of the earlier problems experienced with transistors, AC converters have used power transistors at power ratings up to about 150 kW at 415 V.

The main advantage of transistors is that they can be turned on and off from the *base* terminal, which makes them suitable for *self commutated inverter circuits*. This results in power and control circuits which are simpler than those required for thyristors.

Unfortunately, the base amplification factor of a transistor is fairly low (usually 5 to 10 times) so the trigger circuit of the transistor must be driven by an auxiliary transistor to reduce the magnitude of the base trigger current required from the control circuit. The emitter current from the auxiliary transistor drives the base of the main transistor using the Darlington connection. Figure 3.37 shows a double Darlington connection, but for high power applications, two auxiliary transistors (*triple Darlington*) may be used in cascade to achieve the required amplification factor. The overall amplification factor is approximately the product of the amplification factors of the two (or three) transistors.

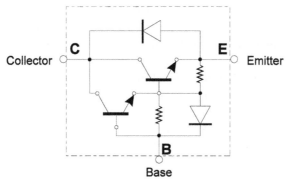

Figure 3.37:
Power Darlington transistor

Transistors, used in VSD applications, are usually manufactured as an integrated circuit and encapsulated into a 3 terminal module, complete with the other necessary components, such as the resistors and anti-parallel protection diode. The module has an insulated base suitable for direct mounting onto the heat-sink. This type of module is sometimes called a *power Darlington transistor module.*

As shown in Figure 3.37, the anti-parallel diode protects the transistors from reverse biasing. In practice, this diode in the integrated construction is slow and may not be fast enough for inverter applications. Consequently, converter manufacturers sometimes use an external fast diode to protect the transistors.

The following figure shows the saturation characteristic of Toshiba MG160 S1UK1 triple Darlington power transistor rated at 1400 V, 160 amp with a built-in free-wheeling diode.

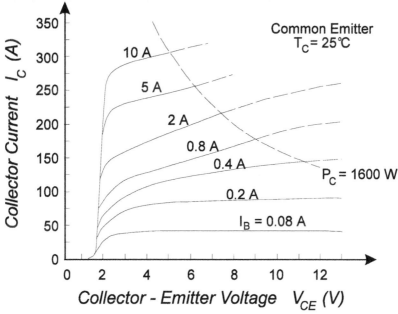

Figure 3.38:
Characteristics of a 160 amp bipolar junction transistor (BJT)

Although the control circuits are completely different, the power circuit performance of a BJT is similar to a GTO thyristor. Forward conduction is blocked until a positive current is applied to the gate terminal and will conduct as long as the voltage is applied.

During forward conduction, it also exhibits a forward voltage drop which causes losses in the power circuit. The BJT may be turned off by applying a negative current to the gate.

The main advantages of the *bipolar junction transistor (BJT)* are:

- Good power handling capabilities
- Low forward conduction voltage drop

The main disadvantages of *BJTs* are:

- Relatively slow switching times
- Inferior safe operating area
- Has complex **current controlled** gate driver requirements

Power bipolar junction transistors are available for ratings up to a maximum of about 300 amps and 1400 volts. For VSDs requiring a higher power rating, GTOs are usually used in the inverter circuit.

3.8.4 Field effect transistor (FET)

A *field effect transistor (FET)* is a special type of transistor that is particularly suitable for high speed switching applications. Its main advantage is that its *gate* is **voltage controlled** rather than **current controlled**. It behaves like a voltage controlled resistance with the capacity for high frequency performance.

FETs are available in a special construction known as the MOSFET. MOS stands for metal oxide silicon. The MOSFET is a three terminal device with terminals called the *source (S)*, *drain (D)*, and the *gate (G)*, corresponding to the emitter, collector, and gate of the NPN transistor.

SYMBOL:

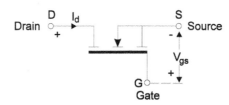

IDEAL: Forward conduction: Resistance (less)
Forward blocking: Loss (less) (no leakage current)
Reverse blocking: Loss (less) (no leakage current)
Switch on/off time: Instantaneous

The overall performance of an FET is similar to a power transistor, except that the gate is **voltage controlled**. Forward conduction is blocked if the gate voltage is low, typically less than 2 volts. When a positive voltage V_{gs} is applied to the gate terminal, the FET conducts and the current will quickly rise in the FET to a level dependent on the gate voltage. The FET will conduct as long as gate voltage is applied. The FET may be turned off by removing the voltage applied to the gate terminal or making it negative.

MOSFETs are majority carrier devices, so they do not suffer from long switching times. With their very short switching times, the switching losses are low. Consequently, they are best suited to high frequency switching applications. A typical performance characteristic of a field effect transistor is shown below.

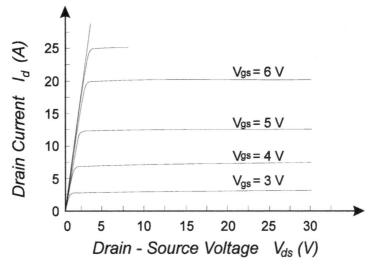

Figure 3.39:
Typical characteristic of a field effect transistor

Initially, high speed switching was not an important requirement for AC converter applications. With the development of *pulse width modulated (PWM)* inverters, high frequency switching has become a desirable feature to provide a smooth output current waveform. Consequently, power FETs were not widely used until recently.

At present, FETs are only used for small PWM frequency converters. Ratings are available from about 100 amp at 50 volt to 5 amp at 1000 volt, but for VSD applications MOSFETs need to be in the 300–600 volt range. The advantages and disadvantages of MOSFETs are almost exactly the opposite of *BJTs*.

The main advantages of a power MOSFET are

- High speed switching capability (10 nsec to 100 nsec)
- Relatively simple protection circuits
- Relatively simple voltage controlled gate driver with low gate current

The main disadvantages of a power MOSFET are

- Relatively low power handling capabilities
- Relatively high forward voltage drop, which results in higher losses than GTOs and BJTs, limits the use of MOSFETs for higher power applications

3.8.5   Insulated gate bipolar transistor (IGBT)

The *insulated gate bipolar transistor* (IGBT) is an attempt to unite the best features of the bipolar junction transistor and the MOSFET technologies. The construction of the IGBT is similar to a MOSFET with an additional layer to provide conductivity modulation, which is the reason for the low conduction voltage of the power BJT.

The IGBT construction avoids the MOSFET's reverse conducting body diode but introduces a parasitic thyristor, which could give spurious operation in early devices. The

IGBT device has good forward blocking but very limited reverse blocking ability. It can operate at higher current densities than either the power BJT or MOSFET allowing a smaller chip size.

The IGBT is a three terminal device. The power terminals are called the *emitter (E)* and *collector (C)*, using the BJT terminology, while the control terminal is called the *gate (G)*, using the MOSFET terminology.

SYMBOL:

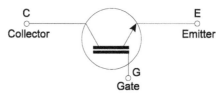

IDEAL: Forward conduction: Resistance (less)
Forward blocking: Loss (less) (no leakage current)
Reverse blocking: Loss (less) (no leakage current)
Switch on/off time: Instantaneous

The electrical equivalent circuit of the IGBT, shown in Figure 3.40, shows that the IGBT can be considered to be a hybrid device, similar to a darlington transistor configuration, with a MOSFET driver and a power bipolar PNP transistor. Although the circuit symbol above suggests that the device is related to a NPN transistor, this should not be taken too literally.

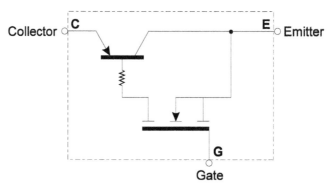

Figure 3.40:
The equivalent circuit of an IGBT

The gate input characteristics and gate drive requirements are very similar to those of a power MOSFET. The threshold voltage is typically 4 V. Turn-on requires 10 V to 15 V and takes about 1 µs. Turn-off takes about 2 µs and can be obtained by applying zero volts to the gate terminal. Turn-off time can be accelerated, when necessary, by using a negative drive voltage. IGBT devices can be produced with faster switching times at the expense of increased forward voltage drop.

An example of a practical IGBT driver circuit is shown in Figure 3.41 below. This circuit can drive two IGBTs, connected to a 1000 V supply, at a switching frequency of 10kHz with propagation times of no more than 1µs.

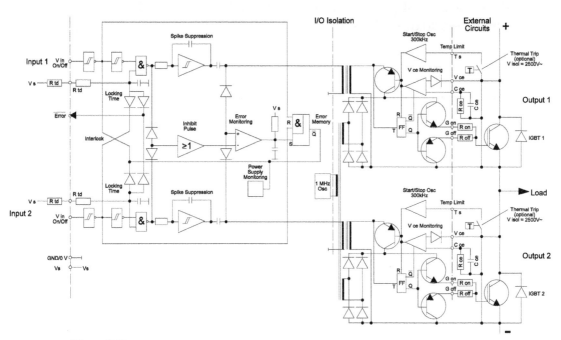

Figure 3.41:
Circuit diagram of semikron SKHI 20 hybrid double IGBT or double MOSFET driver

IGBTs are currently available in ratings from a few amps up to around 500 A at 1500 V, which are suitable for 3-phase AC VSDs rated up to about 500 kW at 380 V/415 V/480 V. They can be used at switching frequencies up to 100kHz. bipolar junction transistors (BJTs) have now largely been replaced by IGBTs for AC variable speed drives.

The main advantages of the insulated gate bipolar transistor (IGBT) are:

- Good power handling capabilities
- Low forward conduction voltage drop of 2 V to 3 V, which is higher than for a BJT but lower than for a MOSFET of similar rating
- This voltage increases with temperature making the device easy to operate in parallel without danger of thermal instability
- High speed switching capability
- Relatively simple voltage controlled gate driver
- Low gate current

Some other important features of the IGBT are:

- There is no secondary breakdown with the IGBT, giving a good *safe operating area* and low switching losses
- Only small snubbers are required
- The inter-electrode capacitances are not as relatively important as in a MOSFET, thus reducing miller feedback
- There is no body diode in the IGBT, as with the MOSFET, and a separate diode must be added in anti-parallel when reverse conduction is required, for example in voltage source inverters

3.8.6 Comparison of power ratings and switching speed of gate controlled power electronic devices

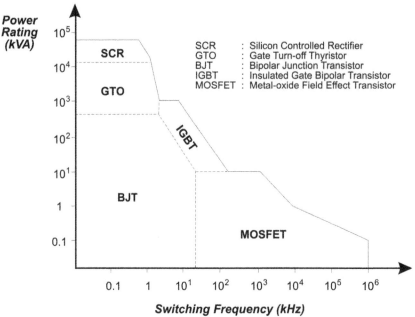

Figure 3.42:
Performance limits of gate controlled devices

3.9 Other power converter circuit components

Inductance

SYMBOL:

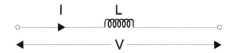

IDEAL: Inductance: constant (linear)
Resistance: zero (no losses)

EQUATIONS: $V = L \dfrac{dI}{dt}$ $\qquad X_L = j\, 2\pi f\, L$

Capacitance

SYMBOL:

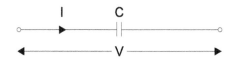

IDEAL: Capacitance: constant (linear)
                   Resistance: infinity (no losses)

EQUATIONS: $I = C\dfrac{\mathrm{d}V}{\mathrm{d}t}$ $X_C = \dfrac{1}{j\,2\pi f\,C}$

Resistance

SYMBOL:

IDEAL: Resistance: constant (linear) and free of inductance and capacitance

EQUATIONS: $V = R\,I$

Transformer

SYMBOL:

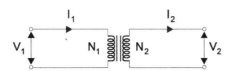

IDEAL: Magnetizing current negligible
                   Free of losses and capacitance

EQUATIONS: $I_1 \times N_1 = I_2 \times N_2$ $\dfrac{V_1}{N_1} = \dfrac{V_2}{N_2}$

4

Electromagnetic compatibility (EMC)

4.1 Introduction

Interference in electrical circuits refers to the presence of unwanted voltages or currents in electrical equipment, which can damage the equipment or degrade its performance.

Electromagnetic interference (EMI) is a fairly broad term that covers a wide range of **undesirable** electrical voltages and currents with a frequency spectrum from DC up to the GHz range. EMI may be introduced into an electric circuit through the following paths:

- Conducted over the power cables or signal cables
- Radiated as an electric or magnetic field from one circuit, which is the source of the interference, and then coupled into another electric circuit, which is the victim

Electromagnetic interference (EMI) includes frequencies in the radio spectrum (100 kHz to 100 MHz) which are known as radio frequency interference (RFI). RFI is the old terminology for the more modern and more general term EMI.

There are two main sources of electromagnetic interference (EMI):

- Natural events such as lightning, electrostatic discharges (ESD) and cosmic discharges
- Man-made interference, which is mainly generated by electrical equipment used for industrial and domestic power supply, communications and control applications

This chapter concentrates on the man-made sources of EMI and mainly those present in the industrial environment. Every electrical circuit should be considered to be a potential source of electrical interference, particularly those where switching of inductive or capacitive circuits takes place. Fortunately, most electrical interference is of a sufficiently low level that it has no noticeable effect on other items of electrical equipment.

Electromagnetic compatibility (EMC) refers to the ability of equipment to function satisfactorily without producing emissions that degrade the performance of other equipment and also are not affected by emissions from other equipment.

Electromagnetic interference (EMI) covers the following main groups:

- Conducted low frequency (LF) interference (up to about 10 kHz)
  - Voltage dips and power interruptions

  - Voltage sags and swells

  - Voltage unbalance

  - Power frequency variation

  - DC in AC circuits and vice versa

  - Harmonics in AC networks (up to approx 3 kHz)

  - Coupled LF voltages and currents

- Radiated low frequency (LF) interference (up to about 10 kHz)
  - LF, electric fields. Radiated from circuits with a high dv/dt

  - LF, magnetic fields. Radiated from circuits with a high di/dt

- Conducted high frequency (HF) interference (from 10 kHz to 1 GHz)
  - Transient over-voltages due to lightning or switching

  - Oscillating transients due to resonance

  - Coupled HF voltages and currents

- Radiated high frequency (HF) interference (from 10k Hz to 1 GHz)
  - HF, electric fields. radiated from circuits with a high dv/dt

  - HF, magnetic fields. radiated from circuits with a high di/dt

The rapid increase in the use of '*non-linear*' *power electronics* devices, such as AC and DC variable speed drives has increased the overall level of electro-magnetic interference (EMI) in industry. To compound the problem, there has been a rapid increase in the number of electronic control and communications devices, which operate at low voltages and high speeds and are susceptible to this high level of interference.

A simple, but effective, way to understand interference problems is to remember that there are always three elements to every interference problem:

- There must be a *source* of interference energy
- There must be a *receptor or victim* that is upset by the interference energy
- There must be a *coupling path* between the *source* and the *receptor*

The management of EMI and EMC in industrial environments falls into two categories:

- The establishment of standards for the containment of EMI by setting maximum limits on EMI emissions from electrical equipment.
- The establishment of standards for the susceptibility (or immunity) of electronic devices through good design and shielding of electronic equipment, which will enable them to operate within certain levels of interference.

A number of EMC standards have been used in industry over the years. In recent times, the international electrotechnical commission (IEC), through technical committee TC77 and its sub-committees, has established the new IEC 1000 series of standards to cover EMC requirements. These standards were introduced in 1996 and have become the basis of EMC standards in a number of countries, including Australia.

Some sections of IEC 1000 are re-issues of earlier IEC standards. For example, sections of IEC 1000 Part-3 replaced the IEC 555 series. Sections of IEC 1000 Part-4 replaced the IEC 801 series. While this re-numbering is an inconvenience in the short term, it will bring the majority of EMC standards into a logical framework, which should facilitate the development of a harmonized set of EMC standards for international use.

The new IEC 1000 series electromagnetic compatibility (EMC) has the following broad structure:

Part-1: General considerations, definitions and terminology
Part-2: The environment
Part-3: Limits for harmonics and voltage variations for equipment connected to AC supplies (replaces IEC 555: 1982)
Part-4: Testing and measurement techniques (replaces IEC 801: 1984)
Part-5: Installation and mitigation guidelines
Part-9: Miscellaneous EMC issues

IEC 1000-1.1 Application and interpretation of definitions and terms
IEC 1000-2.1 Description of environment for LF disturbances
IEC 1000-2.2 Compatibility levels for LF power disturbances
IEC 1000-2.3 Environment - Radiated and conducted phenomena
IEC 1000-2.4 Industrial low frequency conducted disturbances
IEC 1000-3.1 Replaces IEC 555-1: Definitions
IEC 1000-3.2 Replaces IEC 555-2: Harmonics
IEC 1000-3.3 Replaces IEC 555-3: Voltage fluctuations
IEC 1000-4.1 Testing and measurement – Overview of immunity tests
IEC 1000-4.2 Testing and measurement – Electrostatic discharge immunity tests
IEC 1000-4.3 Testing and measurement – Immunity to radiated radio frequency electromagnetic fields
IEC 1000-4.4 Testing and measurement – Electrical fast transient (burst) immunity test
IEC 1000-4.5 Testing and measurement – Surge immunity tests
IEC 1000-4.6 Testing and measurement – Conducted RF disturbance immunity tests
IEC 1000-4.7 Testing and measurement – Harmonic measurement and instrumentation for power supply systems and connected equipment
IEC 1000-4.8 Testing and measurement – Damped power (50 Hz) magnetic field immunity test
IEC 1000-4.9 Testing and measurement – Pulse magnetic field immunity test
IEC 1000-4.10 Testing and measurement – Damped oscillatory magnetic field immunity test
IEC 1000-4.11 Testing and measurement – Voltage dips and short voltage variations immunity tests

4.2   The sources of electromagnetic interference

It is not practical to completely eliminate the electrical interference. The main objective is to minimize their effect on other electronic equipment.

The main sources of EMI in the industrial environment are:

- Any circuit which produces arcs
- Circuits which generate non-sinusoidal voltages, produce electric fields
- Circuits which generate non-sinusoidal currents, produce magnetic fields

AC variable speed drives use power electronic techniques to convert AC to DC (rectifier) and then to convert DC to AC (inverter) to provide a variable voltage variable frequency (VVVF) output. The overall efficiency and performance of the electric motor depends on the quality of the current to the motor. Over the past decade, a smooth sinusoidal current waveform has been achieved through the use of pulse width modulation (PWM) and high frequency switching (10 kHz to 20 kHz). Unfortunately, the AC converter has become a major source of both conducted and radiated electromagnetic interference (EMI).

The two main areas of EMI generation are:

- Supply side (mains)
  The switching frequency of a 6-pulse diode bridge is 300 Hz on a 50 Hz power supply system. The harmonics **generated by the rectifier** fall into the frequency spectrum up to about 3kHz and are conducted back into the power system. The radiated EMI from the rectifier is of relatively low frequency (low di/dt).

- Motor side
  Due to the high inverter switching frequencies (typically between 2 kHz to 20 kHz), high frequency harmonics up to 10 MHz (RFI) are **generated by the inverter** and conducted along the cable to the motor. The EMI radiated from this cable is therefore of relatively high frequency often with high dv/dt.

Supply side harmonic interference is a continuous distortion (up to 3 kHz) of the normal sinusoidal current waveform. The distortion frequencies are multiples of the fundamental 50 Hz frequency.

Harmonic interference comprising mainly low order odd harmonics

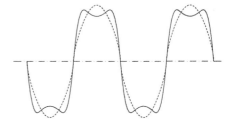

Motor side interference is a continuous high frequency distortion (above about 10 kHz) superimposed on top of the normal sinusoidal waveform.

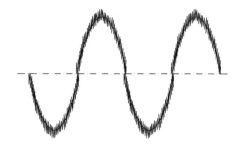

High frequency (RFI) superimposed
on a sinusoidal waveform

AC converters do not themselves radiate a high level of EMI energy. The electromagnetic fields in the immediate vicinity (<100 mm) of the converter can be quite high, but these diminish quite quickly according to the inverse square law and are insignificant at a distance of about 300 mm. When AC converters are mounted in metal enclosures, the electromagnetic radiation is largely eliminated. The main mechanism of propagation of the EMI is through the supply cables, the cables to the motor and most importantly through the earth connections. The supply cable is the most important route for the transfer of EMI. Conduction along the control and communications cables is fairly rare because these cables are usually well shielded and their source impedance is high.

4.3   Harmonics generated on the supply side of AC converters

AC converters use non-linear devices, such as diodes and thyristors, to convert the AC supply voltage to a DC voltage. Rectifiers draw a non-sinusoidal current and distort the AC voltage in the power supply system. They cause additional losses in other items of plant and are the major *source of electromagnetic interference*. Harmonic distortion can be looked upon as a type of electrical pollution in a power system and is of concern because they can affect other connected equipment. As with other types of pollution, the source and magnitude of the harmonic distortion should be clearly understood in order to effectively deal with this problem.

4.3.1   Definitions

The *fundamental frequency* of the AC electric power distribution system is 50 Hz. A *harmonic frequency* is any sinusoidal frequency, which is a multiple of the *fundamental frequency*. Harmonic frequencies can be **even** or **odd multiples** of the sinusoidal fundamental frequency.

The multiple, that the harmonic frequency is of the fundamental frequency, is called the *harmonic order*. Examples of harmonic frequencies of the 50 Hz fundamental are:

Even Harmonics		**Odd Harmonics**	
2nd harmonic	=	100 Hz 3rd harmonic	= 150 Hz
4th harmonic	=	200 Hz 5th harmonic	= 250 Hz
6th harmonic	=	300 Hz 7th harmonic	= 350 Hz
8th harmonic	=	400 Hz 9th harmonic	= 450 Hz
etc		etc	

A **linear** electrical load is one, which draws a purely sinusoidal current when connected to a sinusoidal voltage source, e.g. resistors, capacitors, and inductors. Many of the traditional devices connected to the power distribution system, such as transformers, electric motors and resistive heaters, have linear characteristics.

A **non-linear** electrical load is one, which draws a non-sinusoidal current when connected to a sinusoidal voltage source, e.g. diode bridge, thyristor bridge, etc. Many power electronic devices, such as variable speed drives, rectifiers and UPSs, have non-linear characteristics and result in non-sinusoidal current waveforms or *distorted waveform*. An example of a periodic *distorted waveform*, which repeats itself 50 times a second, is shown in Figure 4.1.

4.3.2   The analysis of the harmonic distortion

The technique used to analyze the level of distortion of a periodic current waveform is known as Fourier analysis. The analysis method is based on the principle that a distorted (non-sinusoidal) periodic waveform is equivalent to, and can be replaced by, the sum of a number of sinusoidal waveforms, which are:

- A sinusoidal waveform at fundamental frequency (50 Hz)
- A number of other sinusoidal waveforms at higher *harmonic frequencies*, which are multiples of the fundamental frequency.

The process of deriving the frequency components of a distorted periodic waveform is achieved mathematically by a technique known as the Fourier transform. Microprocessor based test equipment, which is used for harmonic analysis, can do this very quickly using an on-line technique known as an FFT (fast Fourier transform).

The example below illustrates a distorted voltage wave comprising a fundamental wave and a 3rd order harmonic wave, or simply the *3rd harmonic*, which is a 150 Hz sinusoidal waveform (3 × 50 Hz). The total RMS value of the distorted current is calculated by taking the square root of the sum of the squares of the fundamental and harmonic currents.

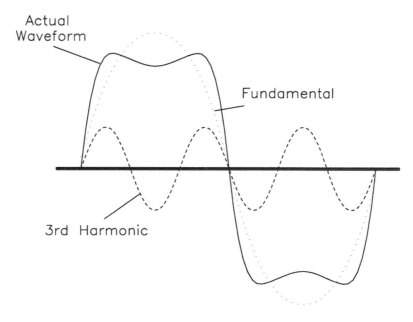

Figure 4.1:
Distorted AC waveform – fundamental plus 3rd harmonic

Harmonic distortion of the current waveform is relatively easy to recognize as a distorted waveform, which is repetitive at the fundamental frequency of 50 Hz. Random noise does not have this repetition. The signature of odd and even harmonics is as follows:

- **Odd harmonics** are present when the negative half cycle is an exact repetition of the positive half cycle, but in the negative direction. Alternatively, odd harmonics are present when the first and third quarters are similar and the second and fourth quarters are similar. Odd harmonics occur with rectifier bridges where the positive and negative half-cycles are symmetrical (even harmonics cancel)
- **Even harmonics** are present when the negative half cycle is NOT a repetition of the positive half cycle. Another characteristic of even harmonics is that the first and fourth quarters are similar and the second and third quarters are similar. It is not common to find even harmonics in an industrial power system.

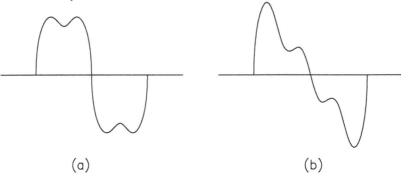

(a)   (b)

Figure 4.2:
Examples of typical distorted AC waveforms

(a) Distorted waveform containing only odd harmonics

(b) Distorted waveform containing only even harmonics

The level of the harmonic distortion generated by VSDs depends on a large number of variables, some of which are often difficult to quantify, such as:

- The magnitude of the current flowing through the converter
- The configuration of the power electronic circuit (6-pulse, 12-pulse, etc)
- The characteristics and impedances of the connected power supply system

The main reason why power electronic converters draw harmonic currents is that the current is *discontinuous* in each phase (refer to Chapter 3). From a harmonics point of view, it does not matter if the rectifier bridge comprises thyristors (controlled rectifier) or diodes (uncontrolled rectifier), they both behave similarly. In the rectifier bridge, only two thyristors (or diodes) are conducting at any one time and the periods of conduction pass from one thyristor (or diode) to the next. Over the period of one cycle of fundamental frequency, each of the 3 phases of the power supply carries a pulse of positive current for a period of 120° and a pulse of negative current for a period of 120°.

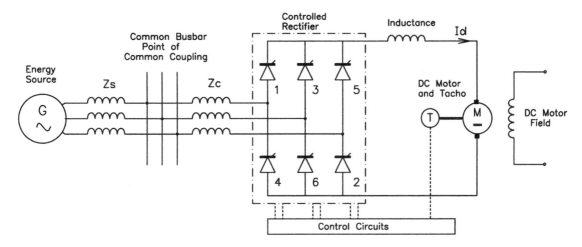

Figure 4.3:
The sources of harmonic currents in a DC converter

These discontinuous phase currents combine on the DC side to result in a rippled DC current, which is usually smoothed by a choke in the DC circuit. Consequently, the rectifier can be considered to be a source of harmonic currents, which flow back into the power supply network impedance.

Power electronic converters do not generate all the possible harmonics, only certain harmonic currents. The harmonic order and magnitude of the harmonic currents generated by any converter depends on 3 main factors:

- The pulse number (p) of the converter. The pulse number is the number of DC pulses produced at the output of the rectifier during one cycle of the supply voltage. The order of the harmonic currents that will be present can be predicted mathematically and is given by the formula:

$$n = k.p \pm 1$$

where: n = order of the harmonics present

k = integers 1, 2, 3,

p = pulse number of the converter

- The magnitude of the load current, I_D on the DC side of the rectifier part of the converter affects the magnitude of the harmonic currents.
- The magnitude of the load voltage, V_D on the DC side of the rectifier part of the converter affects the load current.

4.3.3   Effects of harmonics on other equipment

Harmonic currents cause distortion of the mains voltage waveform that affects the performance of other equipment and creates additional losses and heating. For example, a total harmonic voltage distortion of 2.5% can cause an additional temperature rise of 4°C in induction motors. In cases where resonance can occur between the system capacitance and reactance at harmonic frequencies, voltage distortion can be even higher.

Capacitor banks (used for power factor correction) are particularly vulnerable. They present a low impedance path to high frequency harmonic currents. These increase the dielectric losses in the capacitor bank, which can lead to overloading and eventual failure.

Transformers, motors, cables, busbars and switchgear supplying current to converters should be de-rated (over-dimensioned) to accommodate the additional harmonic currents and the extra losses associated with the high frequency 'skin-effect'. Experience has shown that the current rating of transformers, cables, etc feeding 6-pulse converters must be de-rated by roughly 10% of the converter current and those feeding 12-pulse converters by roughly 5% of the converter rated current.

The electronic equipment used for instrumentation, protection, and control is also affected due to the interference coupled into the equipment or communications cables. This affects the reliability and performance of the control system.

The mains supply current contains currents at the following harmonic frequencies:

$$f_n = (k.p \pm 1) \times f_1$$

where: f_n = frequency of the nth harmonic component of current
f_1 = fundamental frequency of the supply voltage (n = 1)
k = integers 1, 2, 3,
p = pulse number of the connected converter

The following table summarizes the harmonic currents that will be present in the following converter connections.

Converter Connection	Pulse Number p	Order of Harmonics n
1-phase, fullwave	2	3,5,7,9,11,13…
3-phase, halfwave	3	2,4,5,7,8,10…
3-phase, fullwave	6	5,7,11,13,17,19…
Double 3-phase, fullwave one shifted 30°	12	11,13,23,25…

Figure 4.4:
Order of harmonics present for different converter connections

The magnitude of the harmonic currents depend on the active power drawn by the load, which is directly proportional to the DC current I_D. For example, for a 3-phase, 6-pulse converter, the fundamental current is given by:

$$I_1 = \sqrt{3}\,\frac{\sqrt{2}}{\pi}\,I_D = 0.78\,I_D$$

The theoretical magnitude of the harmonic currents can be derived from the following simple formula, based on the assumption that the DC current I_D is completely smooth (ripple-free). In practice, a ripple free DC current is not feasible, so the harmonic currents are invariably larger than the theoretical values.

$$I_n = \frac{I_1}{n}$$

where: I_n the nth harmonic component of current
           I_1 the magnitude of the fundamental component of current
           n order number of the harmonic

For example, the theoretical magnitude of the harmonic currents in the mains, generated by a 3-phase 6-pulse power electronic converter will be:

5th	Harmonic (250 Hz):	20.0% of fundamental current
7th	Harmonic (350 Hz):	14.3% of fundamental current
11th	Harmonic (550 Hz):	9.1% of fundamental current
13th	Harmonic (650 Hz):	7.7% of fundamental current
17th	Harmonic (850 Hz):	5.9% of fundamental current
19th	Harmonic (950 Hz):	5.3% of fundamental current
23rd	Harmonic (1150 Hz):	4.3% of fundamental current
25th	Harmonic (1250 Hz):	4.0% of fundamental current
	etc	etc

The total **RMS current** drawn by a variable speed drive is the square root of the sum of the squares of the harmonic currents.

In a variable speed drive application, assume for example that the current drawn by the 3-phase 6-pulse rectifier at fundamental frequency (50 Hz) is 100 Amps. Using the theoretical values listed above, the following harmonic current values will be flowing:

20 amps (20%) at the 5th harmonic frequency (250 Hz)
14.3 amps (14.3%) at the 7th harmonic frequency (350 Hz)
9.1 amps (9.1%) at the 11th harmonic frequency (550 Hz)
    etc (ignoring harmonics above the 25th harmonic order)

Consequently, the magnitude of the total RMS current drawn by the VSD will be:

$$I_{RMS} = \sqrt{I_1^2 + I_5^2 + I_7^2 + ... + I_{25}^2}$$
$$I_{RMS} = \sqrt{100^2 + 20^2 + 14.3^2 + ... + 4^2}$$
$$I_{RMS} = 104.1 \text{ amps}$$

This illustrates that the total RMS current will be 4.1% greater than value of the fundamental current. This results in extra losses in the cables and transformers that feed the variable speed drive. It is commonly accepted practice to derate the drive cables and transformers by 10%.

These theoretical values are based on ideal commutation and a ripple free load current on the DC link. These ideal conditions do not exist in practice and the magnitude of the harmonic currents depends on several factors, including:

- **Power supply source impedance** – inductance and short-circuit level
- **Inductance of the supply side cables** – are choke fitted
- **Design of the DC link filter** – is a DC link, choke fitted
- **Type of rectifier** – diode bridge or thyristor bridge

The table below illustrates how high the harmonic levels can be without some smoothing. However, this table should be treated with caution, it is aimed at illustrating an example of the '*worst case*' and does not necessarily represent any specific AC converter.

Standards, such as AS 2279 Part 2 gives some typical practical values of harmonic levels which are based on measurement. Reputable manufacturers of VSDs take great care to optimize the design of filters to keep harmonic currents in the supply side as low as possible. On DC drives, the 3 chokes are usually located on the supply side of the converter. This method is seldom used on AC drives, the main technique is to install a choke (inductance) in the DC link. On smaller drives where the actual level of current is quite small, chokes are usually omitted to save space and keep the cost down. This practice has been extended to larger drives by some manufacturers. On AC converters where little or no inductance is used, the level of the harmonic currents can be substantially higher than the theoretical values given in the formula above.

Diode Bridge Rectifier		Harmonic Spectrum			
Circuit Layout	**Phase Current Waveform**	**5th**	**7th**	**11th**	**13th**
- No line choke - No DC link choke - Low source impedance	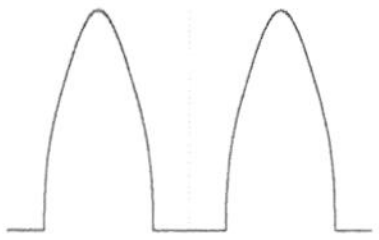	80%	70%	35%	20%
- No line choke - No DC link choke - High source impedance (eg Transformer)	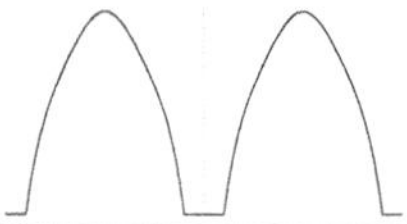	75%	50%	15%	8%
- Line choke fitted - No DC link choke - Low source impedance	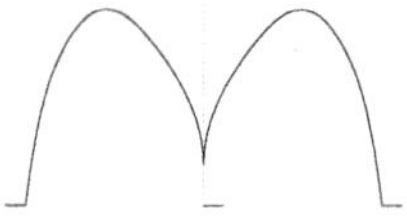	50%	30%	7%	6%
- DC link choke fitted - High source impedance (e.g. Transformer)	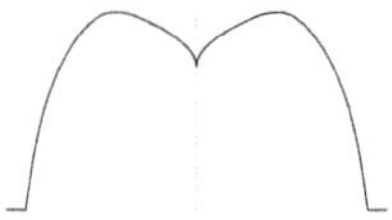	28%	9%	7%	5%

Figure 4.5:
Example of harmonic spectrum with various types of filtering

The mechanical power of the variable speed drive is the product of the output torque and the rotational speed of the motor. This is reflected in the electrical input power that increases with speed.

- For a constant torque load, the active power increases in direct proportion to the speed.

- For a centrifugal fan or pump load, the active power increases as the cube of the speed.

The magnitude and phase angle of the fundamental current, and consequently the harmonic currents, changes as the speed changes. In this respect, PWM converters perform quite differently to DC drives.

In a PWM converter, the DC link voltage remains constant over the entire speed range and is derived from a diode bridge rectifier. As the speed increases, with a constant torque load, the active power increases and, therefore, the DC link current I_D and the RMS value of the fundamental supply current increases in proportion to the speed. The harmonic currents in the supply also increase with speed from an initially low level.

In a DC drive, the DC voltage, which changes in proportion to the speed, is derived from a controlled rectifier bridge. As the speed increases, with a constant torque load, the active power and the DC voltage increase in direct proportion to speed. Therefore, the DC current I_D and the RMS value of the fundamental supply current remains almost constant over the speed range. The DC current I_D and the fundamental current are always slightly higher, compared to the PWM converter, because the firing angle of the controlled rectifier is never zero and the DC voltage is always slightly lower than that of the PWM converter.

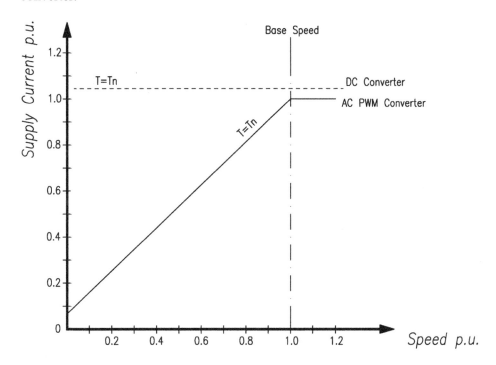

Figure 4.6:
The difference in the supply current drawn by a DC converter and a PWM AC converter of the same capacity at full rated load torque

Figure 4.6 above illustrates these differences between the PWM and DC drives when driving constant torque loads at full rated load torque in the speed range 0 Hz to 50 Hz. With the PWM drive, the harmonic currents decrease with speed reduction because the fundamental current decreases. With the DC drives, the harmonic currents remain roughly constant over the speed range because the fundamental current remains constant. If the load torque is reduced, the converter current will fall in the supply side of both the PWM and DC converter.

The figure below compares the 5th and 7th harmonic currents in an AC PWM drive with the equivalent harmonic currents in a DC drive.

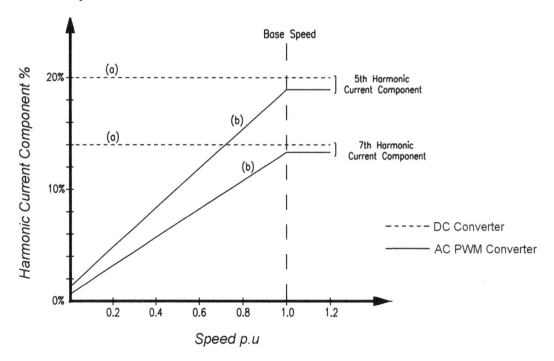

Figure 4.7:
The 5th and 7th harmonic currents at rated torque generated by:

(a) DC converter
(b) PWM-type AC converter

4.3.4   Acceptable levels of distortion in the mains supply system

In the mains supply system, harmonic voltage distortion is the consequence of the flow of harmonic currents through the impedances in the power supply circuit connected to the converter. A typical power supply system at an industrial or mining plant consists of a source of AC power generation, which can either be a local generating station in a small system or a power station at the other end of a transmission line or transformer in a large system. The impedance between the 'ideal' generator and the main busbar is usually referred to as the source impedance Z_s of the supply system. Additional impedance, usually comprising cables, busbars, transformer, etc exists between the main busbar and the converter busbar and is the cable impedance Z_c, as shown in Figure 4.3.

The flow of current to a variable speed motor is controlled by the converter. The current is non-sinusoidal due to the non-linearity of the converter and the generation of harmonic currents. The flow of distorted current through the power distribution and supply system produces a distorted volt drop across the source and distribution impedances in series. Other equipment, such as electric motors or even other consumers can be connected to the main busbar. Consequently, this busbar is referred to as the *point of common coupling* (PCC).

The voltage at the PCC will be distorted to an extent depending on the magnitude of the distorted current, the magnitude of the impedances and the ratio between them. The source impedance can easily be calculated from the system fault level and this is commonly used as the criteria for the permissible size of converter load. A high fault level means a low source impedance and vice versa. If the source impedance is low, then the voltage distortion will be low. The distribution impedance must be calculated from the design details of the distribution system.

A high distribution impedance will tend to reduce the voltage at the point of common coupling but increase it at the converter connection terminals. This voltage distortion can cause interference with the electronic trigger circuits of the converter and give rise to other problems if it becomes too high.

If the magnitude and the frequency of each harmonic current is known, a simple application of Ohm's law will give the magnitude of each harmonic voltage and the sum of them will give the total distorted voltage.

From AS 2279-1991 Part 2, the *total harmonic distortion (THD)* of voltage and current are given by the following formulae. Generally, it is sufficient to use values of *n* up to 25.

$$V_T = \frac{100}{V_1} \sqrt{\sum_{n=2}^{n=\infty} V_n^2} \ \%$$

$$I_T = \frac{100}{I_1} \sqrt{\sum_{n=2}^{n=\infty} I_n^2} \ \%$$

where: V_T = Total harmonic voltage distortion
I_T = Total harmonic current distortion
V_1 = Fundamental voltage at 50 Hz
I_1 = Fundamental current at 50 Hz
V_n = *n*th harmonic voltage
I_n = *n*th harmonic current

The acceptable levels of harmonics in industrial power supply networks are clearly defined in Table 1 of the Australian standard AS 2279-1991 Part 2: *disturbances in mains supply networks*. Briefly, limits are set for the level of total harmonic voltage distortion, which are acceptable at the point of common coupling (PCC). The application of these standards requires the prior calculation of harmonic distortion at all points in the system before the converter equipment can be connected and, under certain circumstances, actual measurements of harmonic voltage to confirm the level of distortion.

4.3.5 Methods of reducing harmonic voltages in the power supply

The use of converters has many technical and economic advantages that will ensure their continued use in industrial and mining plants for many years ahead. In spite of the increase of harmonic distortion in power systems, their advantages far outweigh their disadvantages and their use will continue to grow.

As outlined above, harmonic voltage distortion at the point of common coupling is the result of the flow of harmonic currents through the source impedance. On a *stiff* power system, where the source impedance is low, the voltage distortion will be low. However, the fault level will be high and the short circuit protection equipment will have to be rated accordingly. On a smaller power system, where the source impedance is high, the voltage distortion will tend to be higher.

One of the most practical solutions is to install an inductance (choke) on the supply side of the AC converter to effectively increase the inductive impedance between the converter and the power supply. As shown in the table in Figure 4.5, this effectively reduces the overall level of current distortion, particularly the 5th and 7th current harmonics. The choke can be located internally on the DC link (preferable) or connected externally at the input terminals of the converter. The line chokes need to be of special design to deal with the distorted current waveform. The inductance values of the choke are typically rated between 3% to 5% impedance at fundamental frequency based on the converter rating.

In general, there is not much that can be done to change the source impedance of a power system and, in difficult applications, the solution lies in the techniques to limit the source of the harmonic currents or to divert them to the system earth. There are two main methods of reducing harmonic currents:

The use of multi-pulse converters:

The use of converters of higher pulse numbers will greatly reduce the lower order harmonics. Alternatively, two converters of lower pulse numbers can be combined with a phase shift of $30°$ to produce a system of higher pulse numbers. Theoretically, 12-pulse converters will generate harmonic currents of the order $(12 k \pm 1)$ and will not contain the 5th, 7th, 17th, 19th, etc harmonics. In practice, these do not disappear completely, due to slight differences in converter firing angles and unbalances, but are greatly reduced. The 5th harmonic current usually has the highest magnitude, so its elimination or reduction is desirable. This solution can be expensive.

When several similar converters, with controlled rectifiers, are connected to the same busbar, some cancellation of harmonic currents takes place due to phase shifts between the firing angle of converters running at different speeds. With PWM converters, with diode bridge rectifiers, very little cancellation takes place. The worst case should always be assumed for calculation purposes where the total current for each harmonic is the sum of the currents of the converters operating in parallel.

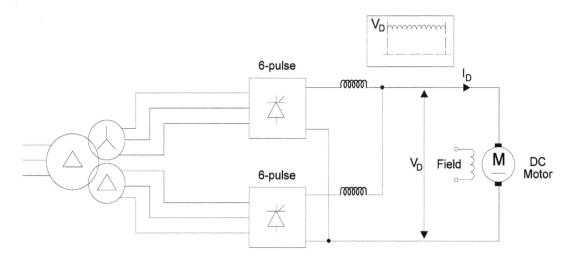

Figure 4.8:
Example of a 12-pulse rectifier bridge feeding a DC drive

The installation of a harmonic line filter close to the converter:
The most common type of harmonic filters used in industry are series L-C filters with some damping resistance. These are usually connected to the busbar (PCC) supplying power to the variable speed drives. Filters may be of relatively simple single-tuned construction, but are usually the more sophisticated (expensive), 2nd or 3rd order filters to provide a wider frequency band. The filter is tuned to specific frequencies so that its impedance is at a minimum at the tuned frequency. The harmonic currents generated by the converter equipment are short-circuited by the filter. The harmonic filter is 'tuned' for a particular frequency when:

$$X_C = X_L$$

or

$$\frac{1}{j\omega C} = j\omega L$$

A typical line side filter comprises resistive, inductive and capacitive components as described below and shown in Figure 4.9.

The main problem with harmonic filters is that they can become detuned over a period of time for any one of the following reasons:

- Changes in the filter capacitance due to age, temperature, or failure of capacitance units within the bank.
- Changes in the inductance due to temperature and current
- Small changes in the system frequency

Since the overall reactance of the filter becomes capacitive at frequencies below the tuned harmonic frequency, resonance can occur between the filter bank and the power system inductance at fundamental or other lower frequencies. This possibility should be considered in the design of harmonic filter equipment to avoid resonance.

4.4 Power factor and displacement factor

When a sinusoidal voltage is connected to a linear load, the result is a sinusoidal current whose magnitude depends on the impedance of the electrical load. The flow of current normally results in power (kW) being consumed in the circuit of the electrical load.

With resistive loads, the current is in phase with the voltage and the total active power is essentially equal to the product of the RMS voltage and the RMS current, which is called the apparent power and is measured as *volt-amperes*. For resistive loads, the ratio between the active power and the total apparent power is equal to 1.

With partially inductive loads, such as electric motors and transformers, the current lags behind the supply voltage by an angle between 0° to 90°, which results in a reduction in the total active power (useful work), which is transferred to the electrical load. The active power consumed is lower than the total *RMS volt-amperes* and the ratio between the active power and the total apparent power falls to a value less than 1.

With purely inductive loads, the current lags behind the supply voltage by 90°, which results in an active power of zero and the ratio between the active power and the total apparent power falls to zero.

The ratio between the active power and the total apparent power is known as the power factor and is defined as follows:

$$Power\ Factor\ =\ \frac{Total\ Average\ Power}{Total\ RMS\ Volt-Amperes}$$

The measurement of power is related to the product of the RMS voltage and RMS current, which is a function of the area under the respective waveforms. With purely sinusoidal voltages and currents, the power factor is a function of the phase displacement angle ϕ between the voltage and the current. Since the phase displacement angle can readily be measured with simple instruments, it is commonly used as a measure of the *power factor*. For purely sinusoidal voltages and currents, the power factor can be shown to be equal to the cosine of the phase displacement angle ϕ.

Cos ϕ is also referred to as the *displacement factor*, which has a value between 0 to 1.

$$Displacement\ Factor\ =\ Cos\ \phi$$

In those cases where both the voltage and the current are purely sinusoidal:

$$Power\ Factor\ =\ Displacement\ Factor\ =\ Cos\ \phi$$

Before the advent of power electronic converters, the power supply voltages, and load currents were sinusoidal and undistorted. The *power factor* was, in general equal to the *displacement factor*, with 1 indicating no lag. This is this quantity, which the manufacturers of AC converters correctly claim as being, '0.95 or better'. However, with non linear power electronic loads, the voltages and currents are distorted and *displacement factor* is not equal to *power factor*.

With non linear loads, with highly distorted currents, the total active power is no longer closely related to the displacement angle between the voltage and current. The harmonic components of the current do not do any useful work and are lost as heat in various parts of the power system and the electrical load. By measuring the total RMS volt-amperes, all these harmonic components are taken into account, which results in a power factor that is much lower than would be calculated from simply measuring the displacement factor. The distorted voltages and currents have to be measured by special true RMS reading instruments, which measure the '*area under the waveform*'. In practice, the real power factor with diode converters can be as low as 0.65, even though the measured displacement factor is greater than 0.95.

The real power factor is also affected by circuit components such as the source impedance of the power system and the inductances in the power electronic circuit. In general, the more distorted the current waveform, the lower the real power factor will be.

Although it is quite easy to measure the power factor of an existing circuit, it is quite difficult to calculate the real power factor of a drive system at the design stage. To achieve an accurate figure in practice, it is necessary to use a computer based circuit analysis program to model the electrical system and take into account the various impedances and the effect of harmonic frequencies on the inductive components of the electrical system.

4.5 Voltages and current on the motor side of PWM inverters

The principles of operation of AC squirrel cage induction motors and the frequency converters to control the speed of these motors are covered in detail in previous chapters and will not be repeated here. This section deals with how the motor responds to the distorted voltages and currents provided at the output terminals of the converter. The DC filter of the converter largely separates the AC input to the rectifier from the AC output from the inverter, so the harmonics on the motor side of the converter may be treated as a separate issue from the harmonics in the supply side.

Users seldom pay much attention to the distorted currents in the motor, apart from applying some minor de-rating factors recommended by the manufacturer of the motor and converter. With the older current source (CSI) and voltage source inverters (VSI), the losses in the motor were significant and it was common practice to de-rate the motor output by as much as 20% to compensate for the harmonic heating in the motor.

With the introduction of inverters with special switching patterns at high switching frequencies, motor currents are almost perfectly sinusoidal and the harmonic losses in the motor are so small that they can usually be ignored. With the thermal margin built into most modern motors, it is now seldom necessary to de-rate the motor for operation with a modern PWM converter.

As described in Chapter 3, most modern AC converters use a voltage source inverter (VSI) to generate a pulse width modulated output voltage. With the introduction of *high frequency switching* above 1 kHz, the harmonics on the motor side are in the frequency spectrums from 10 kHz up to 20 MHz, which is well into the RFI spectrum (>100 kHz). Some of these can pass through the DC link and emerge on the supply side. RFI Filters are now commonly used to prevent this interference being conducted back into the mains. Refer also to Section 4.5.6: RFI Filters.

In contrast to the supply side of the converter, the motor side harmonics are mainly high frequency voltages (high dv/dt), which radiate an electric field. The mathematical analysis of these frequencies is complex and affected by many variables, certainly not as easy as the calculation of supply side harmonics.

The interference generated by the PWM inverter on the motor side and radiated from the motor cable and the converter itself depends on:

- The inverter output frequency range
- The *pulse width modulation (PWM)* switching frequency (typically 2 kHz to 20 kHz)
- The architecture of the inverter, e.g. the internal screening, mechanical details, inductance in motor leads, etc

To comply with the latest EMI standards, it is recommended that the power cable between the converter and the motor be shielded, with the shield connected to earth.

However, shielding can increase the cable shunt capacitance and leakage current.

4.5.1 Effect of the high PWM switching frequency on long motor cables

The high switching frequency of the inverter output voltage interacts with the shunt capacitance of the motor cable, which results in a high frequency leakage current. The higher the leakage current, the higher the losses in the inverter. The leakage current mainly affects the smaller sizes of AC converters (less than 11 kW) because the leakage current is of a similar magnitude as the motor current.

Therefore, modern PWM inverters are designed for a maximum cable length that is determined by the capacitive leakage current losses in the motor cable. Manufacturers can usually provide a de-rating table, which could be similar to the one shown in the Figure below. The de-rating varies for different sizes of converter and also for different manufacturers.

The leakage current is dependent on the length of the cable and its capacitance. This problem is often aggravated by the use of shielded motor cables, which are installed to reduce the radiated EMI from the motor cable. Shielded cables have higher leakage capacitance per meter, almost double that of an unshielded cable. The AC converter needs to be de-rated for long motor cables as shown below.

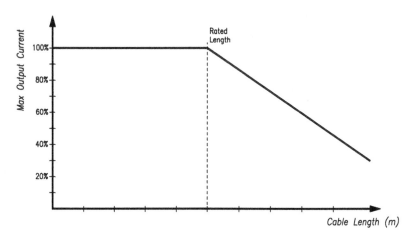

Figure 4.9:
De-rating of the PWM converter for motor cable length

The capacitive leakage current can be reduced by installing a *motor choke (inductance)* at the output terminals of the converter. This series inductance introduces a high impedance between the HF voltage source and the cable capacitance, which reduces the high frequency currents to a relatively low magnitude. These motor chokes are seldom provided as part of the specification of a standard PWM inverter and, where required, are installed as a separate component.

4.5.2   Selection of PWM switching frequency

Many modern AC converters have a selectable output switching frequency and the tendency is to use the highest output frequency to reduce audible noise. The higher the switching frequency, the higher the leakage current losses as described in 4.5.1.

The selection of the PWM switching frequency is a compromise between the losses in the motor and the losses in the inverter.

- When the switching frequency is low, the losses in the motor are higher because the current waveform becomes less sinusoidal
- When the switching frequency increases, motor losses are reduced but the losses in the inverter will increase because of the increased number of commutations. Losses in the motor cable also increase due to the leakage current through the shunt capacitance of the cable.

Manufacturers of converters usually provide a de-rating table or graph, which would be similar to the typical one shown in Figure 4.10.

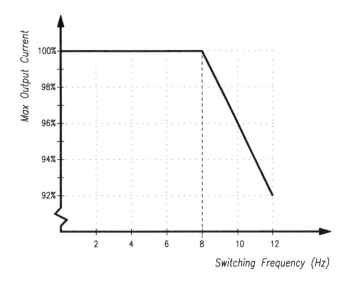

Figure 4.10:
Typical de-rating of the PWM converter for high switching frequency

4.5.3   High rates of rise of voltage (d*v*/d*t*) at inverter output

High frequency switching, using modern IGBT devices in the inverter, achieves a relatively 'smooth' sinusoidal current and reduces the losses in the motor. While the smooth current reduces thermal losses and audible noise in the motor, the sharp rate of rise of the voltage at the inverter output can introduce several other problems. IGBTs have a rate of rise that is several orders of magnitude faster than a BJT. The rate of rise of voltage can be as high as 10 kV/μsec with an IGBT.

Some of the problems that have emerged because of this high switching frequency are:

- High electrical stresses of the cable and motor insulation due to the high rate of rise of voltage (high d*v*/d*t*) and reflections at the end of the motor cable.
- High radiated electric field, due to the high d*v*/d*t*, can exceed the new EMC standards that have been implemented in Europe and Australia.
- As outlined in Section 4.5.1 above, the high d*v*/d*t* across the stray capacitance of cables results in leakage currents which flow into the cable shield (if provided) or alternatively via other conductive paths into the earth. These leakage currents generate additional heat in the inverter or exceed the current limit on smaller converters, which usually results in the converter tripping.

The most significant impact of the high rate of rise of voltage (high d*v*/d*t*) is the high voltage spikes that occur because of the reflected wave at the end of the long motor cable. These voltage spikes can reach peaks of 2 to 2.5 times the inverter DC bus voltage. The phenomenon of reflected waves is quite well understood with communications cables, which operate at similar frequencies. On communications cables the main problem is the interference due to the reflected signal. The doubling of the voltage at the receiving end, due to the reflection, does not cause any physical damage because the signal voltage is usually low.

On modern AC variable speed drives, which use an IGBT inverter bridge, the high voltage spike due to the reflection at the motor end of the cable can damage the insulation of the motor and eventually lead to a short circuit. The mechanism of the failure is as follows:

- The cable between the IGBT inverter output terminals and the AC motor terminals represents an impedance, which comprises resistive, inductive and capacitive components. The cable presents a surge impedance to the voltage pulses generated by the PWM inverter and which travel down the cable. If the surge impedance of the cable does not match the surge impedance of the motor, a partial or full reflection occurs at the motor terminals.
- It is important to understand that this reflection occurs regardless of the type of switching device (IGBT, BJT, MOSFET, GTO, etc) in the inverter. The maximum amplitude of the reflected voltage depends on the velocity of the voltage pulse, its rise time and the length of the cable between the converter and the motor. The rise time of the pulse is related to the switching device. With IGBTs, which have a short rise time (50–500 ns), the length of cable at which **voltage doubling** occurs is much shorter than for a BJT (0.2–2 µs) or a GTO (2–4 µs), which have longer rise times.
- Under worst case conditions, the amplitude of the reflected voltage pulse can be 2 to 2.5 times the inverter DC bus voltage. For a nominal 415 V AC supply voltage to a converter, the DC Bus voltage will be approximately 600 V, which means that the voltage spike at the motor terminals can be as high as 1.5 kV.

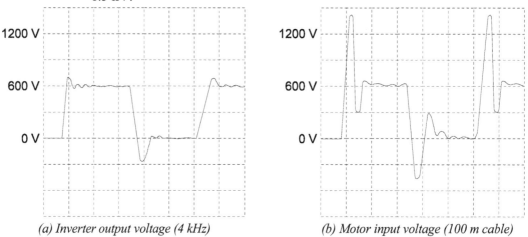

(a) Inverter output voltage (4 kHz)   (b) Motor input voltage (100 m cable)

Figure 4.11:
Comparison of voltage at each end of the motor cable

These voltage spikes could have the following effects on the AC induction motor:

- The first turn of the motor winding is likely to be the worst affected because it is estimated that 60–80% of the voltage spike is likely to be distributed across it.
- The voltage spikes could be greater than the basic insulation level of the electrical phases in the motor, causing dielectric stress of the insulation and eventual failure.
- The voltage spikes could exceed the breakdown voltage of the air separating a winding from the frame at certain points and a partial discharge can occur.

These discharges will degrade the insulation slowly and lead to insulation failure.

- Even if the peak voltage is not high enough to cause a breakdown of the insulation, localized peak capacitive currents heat up the windings. These *hot spots* may aggravate the insulation degradation process. This problem particularly affects older motors, which used insulation materials of a lower temperature rating.

Although the problem of motor insulation stress due to voltage reflections has been present for some time, even with older generations of PWM converters, a few motor insulation failures since the introduction of IGBT inverters has highlighted the problem. On VS drive applications where long motor cables are required, some form of protection should be installed to protect the motor from insulation stress.

While we are not aware of any definitive study that has been done on how these voltage spikes affect the cable insulation, it is felt that the substantial insulation of most cables can withstand these voltage stresses.

4.5.4   Protection of motors against high PWM switching frequency

The use of high frequency PWM switching techniques in modern IGBT inverters has been developed to synthesize a sinusoidal current, thereby reducing the harmonic current losses in the motor and reducing audible noise. These are both desirable features of modern AC variable speed drives.

The new problems due to the leakage currents in the cables and insulation stress in the motor, which have arisen as a result of the high frequency PWM switching techniques, can usually be fairly easily solved.

From point of view of the motor, the best solution is to provide a motor whose insulation can withstand the amplitudes of the reflected voltage spikes. Many motor manufacturers have recognized that motor insulation levels should be increased and have responded with motors that are designed to operate with IGBT inverters. The addition of insulating paper in the motor slots and between phases, can provide additional protection to the most vulnerable areas of the motor windings. This reinforces the benefit of using better quality induction motors for variable speed drive applications.

Since the amplitude of the reflected voltage spike is dependent on the length of the motor cables, these should be kept as short as possible and installations should be planned to minimize the length of motor cables. The table below gives a rough guide to acceptable cable lengths for various inverter types and 415 volt induction motor insulation levels.

Inverter Type	Rated Motor Surge Voltage Withstand Level		
	1,000V Peak	1,200V Peak	1,600V Peak
IGBT (0.1μs)	15m	25m	unlimited
BJT (1μs)	180m	220m	unlimited
GTO (4μs)	700m	850m	unlimited

Figure 4.12:
Recommended maximum motor cable lengths

If motor cable lengths need to be longer than the above recommended values, or when retro-fitting a modern IGBT converter to an existing motor of unknown insulation level, there are a number of solutions which can be used to reduce the effect of the reflected voltage spike on the insulation and thereby extending the life of the drive system.

Output reactors (chokes)

A choke may be installed on the output side of the PWM inverter to increase the inductance of the circuit. While this may introduce a small additional volt drop at motor running frequencies, it also reduces the high rate of rise of voltage, which effectively limits the amplitude of the reflected voltage and extends the permissible length of the motor cable. A choke located at the converter output has the additional advantage that it reduces the leakage current flowing into the capacitance of the cable and reduces the losses in the inverter. Locating the choke at the motor end does nothing to reduce the cable leakage current or the losses in the inverter. Obviously, the insulation of the choke should itself be designed to withstand the high rates of rise of voltage.

Output motor filters

Special harmonic filters, comprising R, L & C components, may also be used in a similar way to the output reactor described above to protect both the cable and the motor. The filter can also be designed to reduce the EMI in the motor cable. The filter achieves this by changing the impedance conditions so that the EMI is diverted into the earth and directed back to the source. The filter mainly comprises a low value series inductance (choke), similar to the choke above, and provides a high impedance to the flow of **high frequency current**, with some additional shunt components. However, the use of shunt capacitance on the inverter side of these filters is restricted due to the effect on inverter performance.

These filters have thermal losses, so the filter losses should be added to the converter losses when determining enclosure cooling requirements. In addition, the filter must be earthed to the same earth bar in the enclosure.

Terminator at the motor terminals

On communications cables, reflected voltages can be attenuated by connecting a *terminator* at the end of the cable. A similar solution can be used with the motor cable. A *terminator*, comprising mainly an *R–C* circuit, connected at the motor terminals can be designed to keep the voltage spike below a potentially destructive level. In comparison to output chokes and filters, *terminators* occupy only a small space, dissipate minimal power and their cost is less than 10% of a filter. In addition, terminators can be used at each motor in multi-motor drive installations.

The following table illustrates the typical maximum motor cable lengths with IGBT converters and the alternative solutions discussed above. The variations in the cable lengths depend on the rated voltage withstand levels of the motor.

Protection System	Maximum Motor Cable Length
No Compensation	10 - 50 metres
Reactor at Inverter	30 - 100 metres
Reactor at Motor	60 - 200 metres
Terminator at Motor	120 - 300 metres

Figure 4.13:
Maximum motor cable lengths with IGBT inverters

4.5.5   Compliance with EMC standards

Various levels of electromagnetic interference (EMI) are generated by all electrical and electronic equipment. EMI is sometimes also referred to as radio frequency interference (RFI). The latter is an 'old-fashioned' term and its continued use is being discouraged in the standards. With the expanded use of variable speed drives (VSDs) throughout industry, the level of EMI generated by this equipment can put at risk the reliable operation of many other electronic devices, such as instrumentation and control devices. However, VSDs are not the only source of EMI, other devices such as fluorescent lamps, switch-mode power supplies, rectifiers, UPS, hand-held radios, mobile phones, etc also generate quite a high level of EMI.

In most industrialized countries, regulating authorities have introduced a framework of EMC standards, which introduce limits for emissions from all electrical/electronic products. At the same time, thresholds of immunity to interference that electrical/electronic products must be able to withstand have also been defined. Products are said to be electromagnetically compatible when they can operate together in the same environment, with limits imposed on those devices that radiate interference and higher levels of immunity for the equipment, which is susceptible being above these limits.

To establish compliance with the EMC framework, manufacturers need to comply with the published standards relevant to the products they supply. In Australia, those items of electrical equipment that comply with the EMC standards can use the compliance mark to signify their compliance. The supplier must take responsibility to ensure that the products comply with the EMC standards. In Europe, the CE mark represents compliance to the similar European standards.

To achieve compliance with the EMC framework in Australia, the supplier must satisfy four basic requirements:

- The supplier must establish sound technical grounds for the product's compliance
- The supplier must make a declaration of conformity
- The supplier must prepare a compliance folder including test reports or a technical construction file
- The supplier must label the product accordingly

From 1st January 1997, products had a 2 year period of grace in which to achieve compliance with the EMC framework. From 1st January 1999, it is mandatory for all electrical products offered for sale (in Australia) in the commercial, residential, and industrial environment to comply with the EMC framework.

The relevant generic standards are as follows:

	Australia	Europe
Generic emission standards	AS/NZS 4251.1	EN50081-1
Generic Immunity standards	AS/NZS 4252.1	EN50082-1

These generic standards call up the tests specified in the relevant IEC-1000 standards.

4.5.6   EMI (or RFI) filters for PWM inverters

When properly designed and used, EMI (RFI) filters connected to the input terminals (line side) of a modern PWM inverter can substantially attenuate the flow of conducted high frequency electromagnetic interference into the power supply cables and into the mains. The best location for the filter is close to the VSD terminals.

In general, a PWM type variable speed drive will not comply with the EMC framework unless it is fitted with a correctly installed RFI filter. Shielding and earthing should be in accordance with the installation instructions supplied with the VSD and/or RFI filter. To achieve EMC compliance, the installation procedure is important. To overcome this dependence on correct installation, many modern VSDs now have the RFI filter built into the VSD as standard equipment.

The line-side filter usually comprises a combination of series inductance and shunt capacitance as shown in Figure 4.14. This filter diverts the harmonic currents away from the power cable and into the local earth connection. Care should be taken to ensure that the earth return cable is installed in such a way that the radiated field does not couple with signal and communications cables.

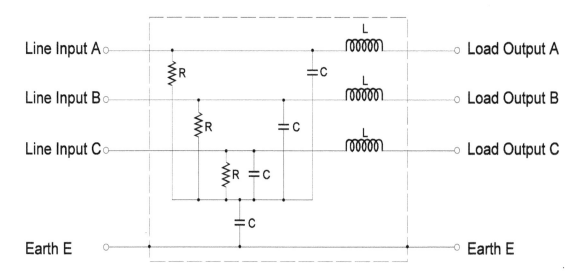

Figure 4.14:
Typical line-side filter for a variable speed drive

4.5.7   Concluding comments about high PWM switching frequency

Although the issue of reflected voltage spikes with IGBT inverters is an important one, clearly there are many drives operating successfully without additional protection. This does not mean that voltage reflections are not taking place, they are below the damage level because either the motor cable is not too long or the cable shunt capacitance is low or the motor insulation level is adequately high. It is not the purpose of this section to over-state the rate of occurrence of this problem. Not all IGBT inverter drive applications will experience a problem. However, users of VSDs should be aware of the potential for this problem to occur and to design the VSD system to minimize its effects.

The following figure summarizes some of the protection features that can be used to improve the harmonic and EMI performance of an AC variable speed drive system.

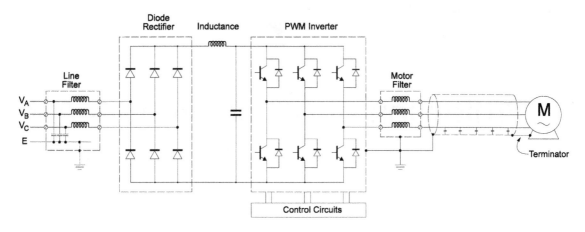

Figure 4.15:
An AC drive fitted with line-side and motor-side filters

5

Protection of AC converters and motors

5.1 Introduction

The protection of AC variable speed drives includes the protection of the following two major components of a VSD:
- The AC converter
- The electric motor

In modern digital AC variable speed drives, most of the protection functions are implemented electronically within the control system of the AC converter. However, to be effective, external sensors are necessary to monitor either the current or the temperature directly. The temperature rise in a motor and converter is the main cause of damage. Since temperature rise is usually the result of high current flow, the sensing of current is a common method of implementing overload and short circuit protection.

5.2 AC frequency converter protection circuits

Digital AC frequency converters usually include a considerable number of protection features to protect the AC converter itself, the output cable and the motor. However, the protection built into the AC converter control system does NOT protect the input side of the converter, which comprises the power supply cable and the rectifier. Short circuit and earth-fault protection must be provided upstream in the main distribution board (DB) or motor control center (MCC). Suitable protection can be provided by:
- Moulded case circuit breakers (MCCB)
- To provide adequate speed, these need to be of the current limiting type
- High rupturing capacity (HRC) fuses

- The fuses are often provided as part of the main isolating switch, which is usually provided for safety isolation. The types of fuses depend on the length of the power supply cable and the inductance of this path.

On the output side, a modern well designed VSD will protect itself from almost all the common faults on the motor side of the converter, such as short-circuit, earth fault, thermal overload, etc. VSDs also usually provide thermal overload protection for the motor.

One of the few faults that will cause damage to the converter is the inadvertent connection of the mains supply to the motor terminals. The inrush through the reverse connected diodes in the inverter will result in inverter damage.

The following protection features are usually available in most modern digital AC converters. These overall protection features are also summarized in Figure 5.3.

- AC input under-voltage protection
- DC bus under-voltage protection
- AC input over-voltage protection
- DC bus over-voltage protection
- Output over-current (short-circuit) protection
- Output earth-fault protection
- Heat-sink over-temperature protection
- Motor thermal over-load protection

5.2.1   AC and DC under-voltage protection

The under-voltage protection system monitors the voltage on the 3 incoming phases as well as the DC bus voltage and responds to various faults as outlined below.

If the supply voltage falls to a low level as a result of some upstream power system fault, it is unlikely that the converter will be damaged. The input diode rectifier of a PWM converter can safely operate at any voltage between zero and the over-voltage trip point. So, a power supply under-voltage event is not really a problem for the power circuit.

Under-voltage protection is mainly required to ensure that all the various power supplies are operating within their required specification. If a power supply unit should lose output voltage regulation, the following could occur:

- The DC bus charging relays may drop out
- The microprocessor could switch to an indeterminate state
- The driver circuits for the main power switches will lose sufficient voltage and current to ensure proper turn-on or turn-off of the switching device
- If there is insufficient turn-on current, a power device may come out of saturation, and attempt to operate in the linear region and losses will increase
- If there is insufficient reverse bias, the power device will be slow to switch off or not switch off at all. Either way, the power electronic switches will fail

Under-voltage protection can be implemented in a number of ways within a VSD:

- Loss of AC supply voltage
  Loss of AC power can be detected by monitoring the three AC line voltages and comparing these to a preset trip point. AC supply under-voltages can be

caused by a complete loss of supply or alternatively a voltage sag (dip) of short duration.

Since the power supply for the converter control circuits is taken from the DC bus via a switch mode power supply (SMPS), it is not necessary to stop the converter immediately the supply voltage is lost. If required, the converter can continue to operate, initially taking energy from the large capacitor on the DC bus. As the DC bus voltage starts to fall, the output frequency can be reduced to allow the motor to behave as an induction generator, driven by the inertia of the mechanical load. This situation could be maintained for a period until the motor stops turning.

Alternatively, the control circuit can be programmed to trip immediately the AC supply voltage is lost. The selection to trip (or not to trip) on loss of AC supply can usually be made by changing a bit in the control logic.

- Loss of DC bus voltage

  The DC voltage can be monitored by a comparator circuit (hardware or software) that compares the DC bus voltage to a preset minimum voltage level. When the DC bus voltage falls below this level, the converter may be shut down (tripped). This trip level is typically set at the lowest rated input voltage, minus 15%. For example, if the VSD is rated at 380 V–460 V ±10%, the lowest specified operating level would be 342 V AC, with an equivalent DC voltage of 485 V DC. The DC bus trip point would typically be set at 485 V DC –15%, that would be 411 V DC.

  In addition to this main DC bus trip point, some of the individual modules sometimes shut down independently. For example, each driver module may have its own under-voltage sensing circuit to ensure that sufficient base or gate drive voltage is available before switching. If these trip, a signal would be returned to the main processor indicating local device failure. These local under-voltage trips are usually used only on critical modules, such as transistor driver circuits.

5.2.2   AC and DC bus over-voltage protection

Ultimately all the electrical components will fail if exposed to a sufficiently high over-voltage. In an AC variable speed drive, over-voltages can occur for the following reasons

- High voltages in the mains power supply
- High voltages generated by the connected motor behaving as an induction generator when trying to reduce the speed of a high inertia load (braking) too quickly

In an AC converter, the DC bus capacitor bank, the DC bus connected power supply module and the main power electronic switching devices have the lowest tolerance to high voltages.

The *capacitor bank* usually consists of individual capacitors in series and parallel. When capacitors are connected in series, the voltage sharing will not be perfect, and so the maximum voltage will be less than the sum of the individual ratings.

For example, if two 400 Vdc capacitors are connected in series, the nominal rating would be 800 V DC. However, the actual safe operating voltage may only be 750 V DC, due to the unequal voltage sharing characteristics. This value will be a function of the capacitor leakage current and the value of the sharing resistor in parallel with each capacitor. The lower the value of the sharing resistor, the better the sharing will be but this will also increase drive losses.

The peak voltage on the DC bus is $\sqrt{2}$ (1.414) times the mains phase voltage. If the maximum rated capacitor voltage is 750 V DC, and allowing for a plus 10% variation in the mains voltage, the practical limit for input voltage is 480 V AC.

The *power semiconductor switching devices*, in the rectifier and inverter, are usually rated at maximum voltage of 1200 V DC. Although this seems well above the maximum capacitor rating, the voltage across a device during turn-off will be much higher than the DC bus voltage, particularly during fault conditions. This is due to stray inductances in the circuit. These voltage peaks can reach about 400 V, so the bus voltage prior to the fault must usually be limited to around 800 V DC maximum, depending on the drive design and the rating of the power devices.

In analog converters, the over-voltage protection is usually a hardware protection system through a simple comparator circuit operating with a fixed set point.

In modern digital converters, the over-voltage protection is usually provided by the microprocessor. This is possible because the DC bus voltage changes relatively slowly, due to the filtering effect of the capacitors.

In microprocessor controlled VSDs, the processor can also provide some over-voltage control. Most DC bus over-voltages are caused by incorrect setting of the deceleration (ramp-down) times of high inertia motor loads. If the deceleration time is set too low compared to the natural run-down time of a rotating load, the motor will behave like an induction generator and power will be transferred from the motor to the DC bus. The DC bus voltage could rise until the over-voltage trip level is reached. Many VSDs have a selectable feature whereby the controller will override the set deceleration time and prevent the over-voltage trip. The DC bus voltage is allowed to rise to a 'safe' high voltage, typically 750 V DC, and rate of deceleration is controlled to keep the voltage below the trip level of 800 V DC.

The under- and over-voltage protection is normally monitored at the DC bus because this is the source of DC power for both the inverter and the control circuits. Typical operating regions and the protection trip levels are summarized in Figure 5.1.

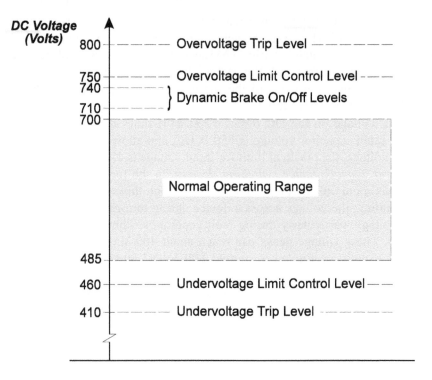

Figure 5.1:
Typical DC bus under- and over-voltage protection levels

5.2.3 Output over-current protection

The purpose of *over-current protection* is to avoid failure of the main power semiconductors (IGBTs, BJTs, MOSFETs, GTOs, etc) during phase-to-phase short circuits on the motor side of the converter. Unlike diodes and SCRs, fuses are not appropriate for the protection of most power switches due to their I^2t characteristics.

The most effective method of protection is to switch all the inverter switching devices off when the current rises above a given set point. The protection level is dependent on their safe operating area characteristic. This maximum fault current is usually what determines the maximum rating of the drive. Typically, the trip current is around 200% of the drive current rating, with current limiting at 150% or sometimes 180%.

To maximize the effective rating of the VSD, it may be possible to operate closer to the trip current if the *rate of rise* (di/dt) of current is controlled. This can be achieved by introducing a choke between the power semiconductor device and the output terminals of the VSD. If a short circuit occurs on the VSD output, the rate of change of current (di/dt) will be equal to the bus voltage divided by the inductance:

$$\frac{\mathrm{d}i}{\mathrm{d}t} = \frac{V_{bus}}{L_{out}}$$

For example, with a 600 V DC bus voltage and a 100 μH output choke, the current rise time will be limited to 6 amp/μsec. For a short circuit on the output of a 50 kW AC converter, with a trip current level of 200 amp, it will take 33.3 μsec to reach the trip point.

This is significant when considering the *propagation delay* through the current feedback and protection circuits. The propagation delay is the amount of time between the actual current reaching the trip point and the turn off of the power devices. This delay exists in the current measuring device, the amplifiers through which the signal passes, the comparator itself, the power device driver circuit and the actual power device.

If the propagation delay and the rate of change of current are known, then the actual device current when the power devices switch off can be estimated. For example, if the total propagation delay is 3 μsec and the di/dt is 6 amp/μS, then the actual device current will be 18 amps higher than the current trip point when the devices actually turn off. This is called *current overshoot*.

While larger output chokes will reduce this overshoot and have a few other advantages, they also introduce losses, are bulky and expensive. For this reason it is important to minimize the propagation delay in the over-current protection circuit. As a result, high bandwidth current feedback and amplifiers are usually used. To minimize propagation delays in the microprocessor, it is common for over-current protection to be performed completely in hardware, even in a digital VSD.

Over-current events can also occur as a result of sudden increases in the load torque on the motor or during periods of high motor acceleration. These increases in current occur relatively slowly, allowing the current to be monitored and controlled by the microprocessor. The increase in current can be limited to a preset value typically of up to 150% of the rated converter current. The current limit control system regulates the output frequency in such a way that it reduces the motor torque. If the over-current is due to a high rate of acceleration, current is reduced by reducing the rate of increase of current. If the over-current is due to a temporary motor overload, the output speed may be reduced.

Typical over-current protection and current limit levels are summarized in Figure 5.2.

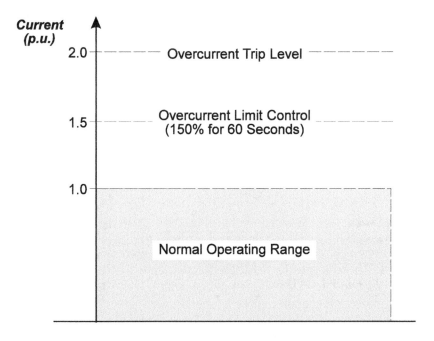

Figure 5.2:
Typical over-current protection levels and current limit settings

5.2.4   Output earth fault protection

Earth fault protection is designed to detect a short circuit between a phase and earth, on the output side of the VSD, and immediately shuts down the converter. This protection is generally not intended for protection of human life from electric shock, as the trip points are set much higher than acceptable human safety limits. This feature is mainly for the protection of the AC converter itself.

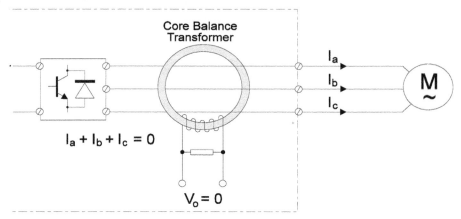

Figure 5.3:
Core balance current transformer for earth fault protection. The normal operating condition, no earth fault present

Earth fault protection is usually implemented by means of a *core balance current transformer*. This is constructed from a toroidal magnetic core through which either the DC bus cables or the output motor phase cables are passed. A low current secondary winding is wound around the toroid and connected to the protection circuit. If the vector sum of all the currents passing through the core add up to zero, the flux in the core will be zero. A net zero flux is the normal operating situation.

If an earth fault occurs and there is a path for current to earth, the sum of the currents through the core balance transformer will no longer be zero and there will be a flux in the core as shown in Figure 5.4.

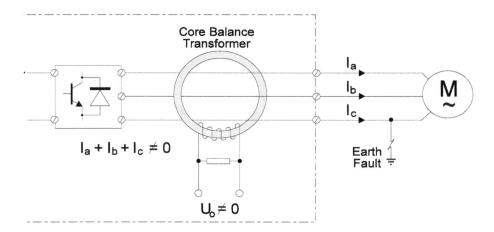

Figure 5.4:
Core balance current transformer for earth fault protection. Earth fault condition, the net current is not equal to zero

This flux will result in a current being generated in the secondary protection winding, which is converted to a voltage via a burden resistor. A comparator circuit detects the fault and shuts down all the power device drives. Typically, the protection trip level is around 5 amp.

Care must be taken in establishing the set point for the earth fault trip circuit. In all PWM VSDs, some leakage current will always take place to earth due to the high frequency components of the motor current waveform and the capacitance of the motor cables to earth. High leakage currents can sometimes cause some nuisance tripping of the earth fault protection.

5.2.5   Heat-sink over-temperature protection

Over-temperature protection is usually provided to prevent over heating of various components in the converter, particularly the junction temperature of the power semiconductors, which is limited to 150°C. To ensure this limit is not reached, the heat-sink temperatures are usually maintained at temperatures below 80°C to 90°C, depending on the actual design. Consequently most heat-sinks are fitted with temperature sensors or switches to detect when the maximum temperatures are reached.

Other modules, such as the power supplies or device driver modules, may have their own individual over-temperature protection. It is common to measure ambient air temperature close to the control electronics to ensure this does not exceed device ratings (usually ±70°C).

Low cost drives may rely on simple bimetallic temperature switches (microtherms), which operate at a specific temperature. However, most modern drives use *silicon junction temperature sensors* to feed back the actual temperature to the microprocessor.

Using this method, the processor can provide a warning to the operator prior to actual shutdown. On more advanced VSDs, some corrective action might be taken automatically, such as reducing the motor speed or reducing the PWM switching frequency.

5.2.6   Motor thermal overload protection

Almost all modern VSDs include some provision for motor thermal overload protection.

The simplest form of protection is to make provision for a digital input, which shuts down the drive when some external device, such as a thermal overload or thermistor relay is activated. Many manufacturers of VSD now make provision for a direct input from a thermistor sensor, so that only the thermistors need be placed in the motor windings and eliminates the need for a thermistor relay. The inputs are normally delivered with a resistor connected across the terminals, which should be removed during commissioning. This often creates some difficulties during commissioning for those who do not read the installation manuals.

The most common method used for motor thermal overload protection on digital VSDs is to use the **current sensing method** with a *motor protection model* as part of the microprocessor control program. The measurement of motor current is necessary for other purposes, so it is a small step to provide motor thermal modeling. The model can continuously estimate the thermal conditions in the motor and shuts down the VSD if limits are exceeded.

The simplest motor model is to simulate a eutectic thermal overload relay by integrating motor current over time. This simplistic method does not provide good motor protection because the cooling and heating time constants of the motor change at different speeds.

Over a period of time, the motor protection features in VSDs have become more sophisticated by using motor frequency as an input so that shaft fan cooling performance, at various speeds, can also be modeled. The most advanced VSDs require motor parameters such as rated speed, current, voltage, power factor and power to be entered to enable a comprehensive motor thermal model to be implemented in software, affording excellent motor protection without any direct temperature measurement devices.

For these motor models to be accurate and effective, previous conditions need to be stored in a non-volatile memory in case the power is interrupted. This can be achieved through simple devices such as capacitors or non-volatile memory chips, such as EEPROMs and NVPROMs.

5.2.7   Overall protection and diagnostics

The following block diagram is a summary of the protection features commonly used in modern digital PWM AC converters. As outlined above, many of these protection functions are implemented in software, using suitable algorithms. The main exceptions are the over-current protection and the earth fault protection, which are implemented in hardware to ensure that they be sufficiently fast to adequately protect the power semiconductor devices.

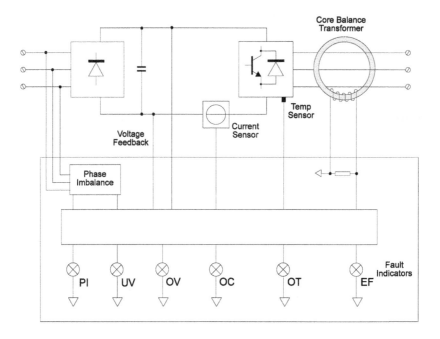

Figure 5.5:
Example of VSD protection block diagram

5.3   Operator information and fault diagnostics

Modern digital variable speed drives (VSDs) all have some form of *operator interface module* which provides access to internal data about control and status parameters during normal operation and diagnostic information during fault conditions. This module is sometimes called the *human interface module* (HIM), or something similar. The HIM usually provides an LED or LCD display and some buttons to interrogate the control circuit. This operator interface can also be used to install and change VSD settings parameters.

In addition, modern VSDs also permit the transfer of these parameters to remote locations via *serial digital data communications*. Some details about the serial communication are covered in the section on installation in Chapter 8. The communications interface permits control from a remote *programmable logic controller (PLC)* as part of an overall automated control system. The diagnostic information can be transferred over the serial interface to a central control center so that an operator can take full advantage of the information available.

When an internal or external fault occurs, the control circuit registers the type of fault, which helps to identify the cause of the fault and the subsequent rectification of the problem. Modern microprocessor controlled converters employ a diagnostic system that monitors both the internal and external operating conditions and responds to any faults in the way programmed by the user. The control system retains the fault information in a non-volatile memory for later analysis of the events that occurred. This feature is known as *fault diagnostics*.

There are three main levels of operator information and fault diagnostics:

- The **first level** provides information about the on-going situation inside a VSD and refers mainly to the setting parameters and the real-time operating parameters and metering information, such as output voltage, output current, output frequency, etc.
- The **second level** provides diagnostic information about the status of the protection circuits and will indicate the *external faults* as described above.
- The **third level** provides diagnostic information about the status of *internal faults*, such as the identification of failed modules. Dedicated internal diagnostics are usually only found in high performance VSDs.

The following is a brief list of typical internal parameters and fault conditions.

Module	Parameters and fault diagnostics
Power supply	Power supply voltage, current and frequency
DC bus	DC link voltage and current
Motor	Output voltage, current, frequency, speed, torque, temperature
Control signals	Setpoint, process variable, error, ramp times
Status	Protection circuits, module failures, internal temps, fans running, switching frequency, current limit, motor protection, etc
Fault conditions	Power device fault, power supply failed, driver circuit failed, current feedback failed, voltage feedback failed, main controller failed

Figure 5.6:
Typical list of variable speed drive parameters

At the **first level**, most modern digital VSDs provide information about the status of:
- All setting parameters which define the operating conditions
- The digital inputs and outputs, such as start, stop, enabled, jog, forward/reverse, etc
- The status of analog inputs, such as speed reference, torque reference, etc

• The real-time operating parameters, which include a vast array of information, such as output frequency, output voltage, output current, etc

At the **second level**, when a fault occurs and the VSD stops, diagnostic information is provided to assist in the rectification of the fault, thereby reducing downtime. There is always some overlap between these levels of diagnostics. For example a persistent over-current trip with no motor connected can indicate a failed power electronic switching device inside the converter.

The table in Figure 5.7 shows the most common *external* fault indications provided by the VSD diagnostics system and the possible internal or external problems that may have caused them.

Protection	Internal Fault	External Fault
Over-voltage	Deceleration rate set too fast	Mains voltage too high transient over-voltage spike
Under-voltage	Internal power supply failed	Mains voltage too low Voltage sag present
Over-current	Power electronic switch failed driver circuit failed	Short circuit in motor or cable
Thermal overload	Control circuit failed	Motor over-loaded or stalled
Earth fault	Internal earth fault	Earth fault in motor or cable
Over-temperature	Cooling fan failed heat-sink blocked	Ambient too high enclosure cooling blocked
Thermistor trip		Motor thermistor protection

Figure 5.7:
Variable speed drive diagnostics table

The *internal diagnostics* system can provide an operator with information about faults that have occurred inside the drive. This can be further broken down into fault conditions, such as a failed output device, commutation failures, etc. Fault conditions are indications that a particular module or device has failed or is not operating normally. To provide fault condition monitoring, the drive must be specifically designed to include internal fault diagnostic circuits.

For example, power semiconductor drivers may include circuits that measure the *saturation voltage*, which is the voltage across the device when it is on, for each power semiconductor. This can identify a short circuit in the power switch and the VSD can be shut down before the external over-current trip or fuses can operate.

Considerable cost and effort is required to implement internal fault condition monitoring, and only a few high performance VSDs provide extensive internal diagnostics. This feature can be very useful for trouble-shooting, but this is usually only warranted when down time represents a major cost to the user.

5.4 Electric motor protection

The useful life of an electric motor is dependent on the following main components:

• **Electrical parts**, such as the stator windings & insulation, the rotor windings & insulation and their respective external connections

- **Mechanical parts**, such as the stator core with slots, the rotor core with slots, the shaft, the bearings, the frame & end shields and the cooling system.

Using modern materials, most of these components can be designed and constructed to have a high level of reliability. Experience has shown that mechanical failure is rare and the most likely causes of failure are:

- Motor overloading, current exceeds rated level for a period of time
- Frequent starting, inching, jogging & reverse plugging, high currents
- Single phasing or unbalanced power supply, high currents
- Stalling, high currents
- High ambient temperature
- Loss of cooling

During the above abnormal operating conditions, the temperatures in the stator and/or the rotor windings can rise to excessive levels, which causes the degradation of the insulation materials used to isolate the windings from each other and the earthed frame of the motor.

The temperature rise in a motor winding is mainly due to the I^2R losses, or copper losses, where the heat is generated by the load current (I) flowing through the resistance (R) of the stator windings. During design, the cross-sectional area of the stator windings is selected with a particular maximum load current in mind. The design objective is to balance the I^2R losses, at maximum rated load, with adequate ventilation or cooling so that the resulting temperature rise in the winding will be below the critical temperature of the insulation materials chosen. In AC motors, the stator current is proportional to the mechanical load torque. In DC motors, the armature current is proportional to the mechanical load torque. Consequently, each standard motor size is rated for a maximum stator or armature current.

Excessive winding temperature most commonly occurs when the load current exceeds the maximum rated value. This condition is called *thermal overloading*.

When the temperature in a winding rises above a certain critical level, the insulation is permanently damaged. The critical temperature, above which permanent damage takes place, is dependent on the type of insulation material used. In the standards, the different types of insulation are classified into Classes, such as Class-B, Class-F, Class-H, etc.

For example, the temperature in a winding with Class-F insulation is permitted to safely rise to a maximum of 140°C, or 100°C above the commonly specified maximum ambient temperature of 40°C, without permanent damage to the insulation.

If the working temperature of the winding increases above 140°C, the characteristics of the insulation material start to degrade. Above 155°C, the insulation will be permanently damaged and its useful life sharply reduced. Insulation failure results in short circuits or earth faults, which would require the replacement or repair of the faulted winding. Long insulation life is particularly important for electric motors, which operate in strategic locations in industry under continuously changing load conditions. A constantly applied temperature rise of just 10°C above the maximum rated temperature can reduce the useful life of a motor to 50% of its original value, as illustrated in the curve below.

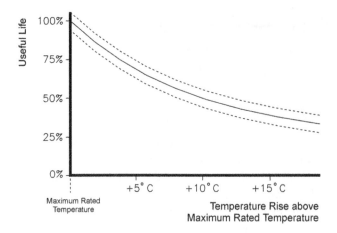

Figure 5.8:
The effect of temperature rise above maximum rated temperature on the useful life of an electric motor

To protect a motor from insulation damage due to excessive temperatures, any potentially damaging operating condition should be detected by a sensing device and the motor should be disconnected from the power supply before insulation damage can occur. The most common devices used for the protection of electric motors are:

- **Current sensing devices**, such as thermal overload protection relays, which continuously monitor the primary current flowing into the motor windings and initiate a trip when a preset current level is exceeded.
- **Direct temperature sensing devices**, such as thermostats, thermistors, thermocouples and RTDs, which continuously monitor the actual temperature in the motor windings and initiate a trip when a preset temperature level is exceeded.

5.5   Thermal overload protection – current sensors

Current sensing thermal overload (TOL) protection relays, whether of the indirectly heated bimetallic type or the electronic type, monitor the stator current in AC motors or armature current in DC motors, and use this information to determine if the motor has become over-loaded. TOL relays should be designed to match the thermal characteristics of the motor.

On smaller motors, a bimetallic type of TOL relay is normally mounted in conjunction with the motor contactor, which opens when an overload condition is detected. Additional features usually include phase failure (single phasing) detection. The main advantage of the bimetallic TOL relay is its low cost and simplicity.

Bimetallic TOL relays do not provide adequate protection for repeated starting, jogging and other periodic duties. The reason is that the heating and cooling time-constant of the bimetal are equal, whereas the cooling time-constant of a typical squirrel cage motor is approximately twice its heating time-constant, mainly because the cooling fan stops when the motor is stationary. During repeated starting and jogging, the bimetal cools down faster than the motor and, consequently, does not provide adequate thermal protection.

For larger motors and those with an intermittent duty, it is necessary to use an electronic motor protection relay, whose thermal characteristics and settings are designed to more closely match those of the motor. In this case, overload protection is usually part of an overall motor protection relay, which also provides protection against short circuits, earth faults, stalling, single phasing, multiple starts, etc. Several adjustable settings enable the motor protection relay to be matched to the type, size and application of the motor. Modern microprocessor relays can also store and display data such as line currents, unbalance currents, thermal capacity of the motor, etc, and transfer this information to a remote host computer over a serial communications link.

Although *current sensing* TOL protection devices, which monitor stator or armature current, are cost effective and have a reasonably good response time, they seldom take into account other environmental conditions, such as reduced cooling, restricted or total loss of ventilation or excessively high ambient temperatures. For example, in AC variable speed drives, the shaft mounted fan cooling on a standard AC motor is reduced as the motor speed is reduced, which changes the heating time constant of the motor when it is running at speeds below 50 Hz. Most modern digital AC converters have built-in thermal overload protection, which is designed to compensate for the changes in the heating and cooling time constants as the speed is adjusted. But, monitoring the stator current is not always a reliable method for protecting the motor winding insulation from damage due to over-temperature.

5.6 Thermal overload protection – direct temperature sensing

For the more difficult applications, *direct temperature sensing* of the winding temperature at *hot spots* or other strategic points is preferable. There are several types of devices that can be used for *direct temperature sensing*. Some of the most common techniques are summarized in the table in Figure 5.9.

The following are some of the applications where **direct temperature sensing** is considered to be more reliable than stator or armature **current sensing**:

- AC squirrel cage induction motors supplied from AC frequency converters
- AC motors which have frequent transient overloads
- AC motors which are frequently stopped or started
- AC motors in high inertia applications with long starting times
- AC motors in applications where the rotor can lock or stall
- DC motors controlled from DC converters
- Thermal protection in mechanical applications, such as large bearings, gear housings, oil baths, heat sinks, etc

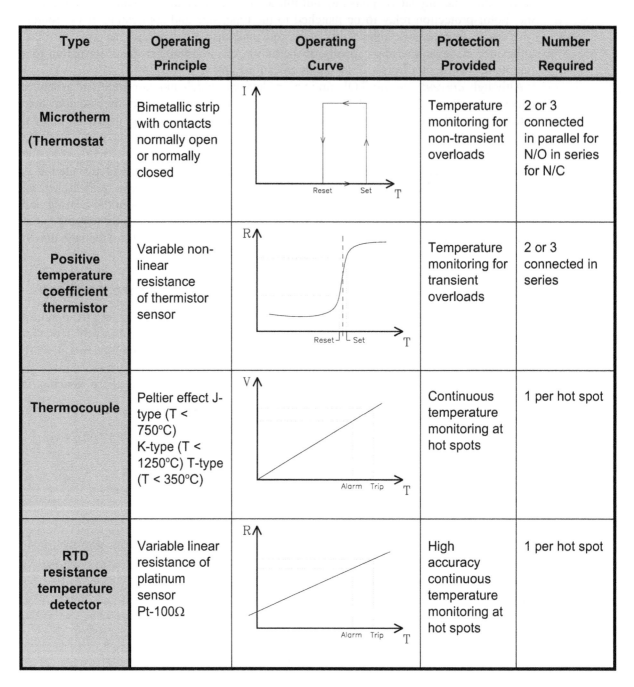

Type	Operating Principle	Operating Curve	Protection Provided	Number Required
Microtherm (Thermostat	Bimetallic strip with contacts normally open or normally closed		Temperature monitoring for non-transient overloads	2 or 3 connected in parallel for N/O in series for N/C
Positive temperature coefficient thermistor	Variable non-linear resistance of thermistor sensor		Temperature monitoring for transient overloads	2 or 3 connected in series
Thermocouple	Peltier effect J-type (T < 750°C) K-type (T < 1250°C) T-type (T < 350°C)		Continuous temperature monitoring at hot spots	1 per hot spot
RTD resistance temperature detector	Variable linear resistance of platinum sensor Pt-100Ω		High accuracy continuous temperature monitoring at hot spots	1 per hot spot

Figure 5.9:
Protection devices used for direct temperature measurement

In practical applications, one or more *direct measuring thermal sensors* are usually used to monitor the temperature at several strategic points in an electric motor. These sensors are used in conjunction with an associated relay or controller, which is connected in the motor control circuit to provide the following:

- **Alarm**: Draws the attention of the operator to the high temperature condition, using audible and/or visual alarms, without tripping the motor
- **Trip**: Stops the motor by tripping the power supply circuit to the motor

To achieve the objectives of separate alarm and trip setpoints, microtherms and thermistors require a group of two sensors at each strategic point. The first, with a lower temperature setpoint, is used to provide the alarm function and the second, with a higher temperature setpoint, is used to provide a trip function.

With thermocouples and RTDs, which can continuously measure the actual temperature at each strategic point, the electronic controller normally has two preset temperature levels for alarm and trip. Two separate contact outputs can then be used to initiate and alarm or trip the motor.

A detailed description of the various direct temperature sensing methods of motor protection is given in Appendix A: Motor protection – direct temperature sensing.

6

Control systems for AC variable speed drives

6.1　The overall control system

Most modern AC variable speed drives (VSDs) are of modular construction. Some of the technical details of the main components, such as the input rectifier, DC link, output inverter and the connected motor have already been covered in the previous chapters. This chapter covers the control system, embodied in the control circuits.

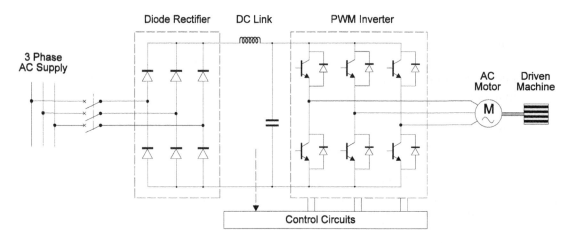

Figure 6.1:
Main components of an AC variable speed drive

Although the main function of the control system for modern PWM-type AC VVVF converters is to control the semiconductor switches of the PWM inverter, there are a number of other important functions, which need to be controlled. The overall control system can be divided into 4 main areas:

- Inverter control system
- Speed feedback and control system
- Current feedback and control system
- External interface, which includes the following:
  - Parameter settings by the user
  - Operator information and fault diagnostics
  - Digital and analog inputs to receive control signals (start, stop, etc)
  - Digital and analog outputs to pass on status information (running, faulted, etc)

With the rapid advances is digital electronics over the last decade, modern VSD control systems are based on one or more microprocessors. The control system must be designed to achieve the following main objectives:

- High level of **reliability**
- High **inverter performance** to ensure that the output current waveform provides sufficient motor torque, at selected speed, with minimum of motor losses
- **Inverter losses** should be minimized
- It must be possible to **integrate the control system into the overall process control system**, with facilities for external control and communications interfaces
- High tolerance to power supply fluctuations and EMI

6.2   Power supply to the control system

For reliable operation of a VSD, it is essential that a reliable power supply is available to provide power to the control circuits of the AC converter, even under abnormal situations, such as a power dip, high levels of interference, etc. The general requirements for power in a modern VSD are set out in the table below.

Functional block	Approximate load	Common potential	Voltages
Control circuits	20 VA independent of drive size	Usually referenced to earth	±5 V for µP ±12 V or ±15 V
Power interface circuits	20 VA, depends on current feedback method	Usually referenced to earth, may have auxiliary isolated supply at bus potential	±12 V or ±15 V
User interface circuits	5 VA for logic inputs 20 VA for opto inputs	Earth referenced if using logic inputs. Isolated (floating) if using opto inputs	5 V or 12 V if logic inputs 12 V or 24 V if opto inputs

Power device drivers	0.2 VA–1 VA per kW depends on power device highest for GTO less for BJT lowest for FET/IGBT	4 or 6 separate isolated outputs, 3 at motor phase potential, 1 or 3 at –ve DC bus potential	±12 V, 15 V, 24 V Turn on, plus Turn off, minus wrt power device
DC bus charging circuits	0.1 VA per drive kW	Earth referenced or floating	12 V DC, 24 V DC 110 V, 240 V or 415 V AC
Cooling fan power	0.5 VA per drive kW	Earth referenced or floating	12 V DC, 24 V DC 110 V, 240 V or 415 V AC

Figure 6.2:
General requirements for power in a PWM variable speed drive

The simplest method of providing auxiliary power to the converter control circuits is from an auxiliary transformer connected to the mains. Multiple secondary windings are necessary to provide isolation for the control circuits and the device drivers. The major problem with this approach arises when there is an interruption of the mains power.

Control of the inverter is lost and the VSD would have to be stopped, even for short dips in the supply. In many drive applications, there is a requirement for VSDs to '*ride through*' voltage dips of short duration.

Consequently, most modern AC converters use *switched mode power supplies* (SMPS), fed directly from the DC link, to provide the auxiliary power to the control system. These are essentially DC–DC converter. The main advantage of this approach is that control power can be maintained right up to the time that the motor stops, irrespective of the condition of the mains supply. When the mains power fails, auxiliary power is maintained initially from the large capacitors connected across the DC link and later from the inertia motor itself. When mains power is interrupted, most AC converters are programmed to reduce frequency and retrieve power from the motor, which behaves as an AC induction generator when the frequency is reduced.

There are many types of switched mode power supplies, including *fly-back* converters, *forward* converters and *bridge* converters. They can be isolated or non-isolated and have single or multiple outputs. Since they operate at high frequency (10 kHz to 100 kHz), they are physically much smaller than conventional mains frequency transformer based power supplies and despite the added complexity of SMPSs, they are of comparable cost.

Due to the modular nature of modern drives, it is common to have multiple auxiliary power supplies, each of which is dedicated to a single module of the VSD, such as the control module, the pulse amplifier driver stage, the cooling fans, etc. These different SMPSs may operate independently from the DC link or from a central SMPS that converts the DC link voltage to a single isolated low voltage supply, such as 24 V DC. Each module may then take its power requirements from this 24 V DC power supply.

As shown in table of Figure 6.2, the device driver power supplies need to be provided with 4 or 6 isolated power outputs. These need to be isolated because the three power electronic switches connected to the positive terminal of the DC link have their emitter

(IGBT & BJT), source (MOSFET) or cathode (GTO) terminals connected to the output phases to the motor. This terminal is the reference terminal for the driver stage, while the base or gate terminal must be driven **positive to turn on** or **negative to turn off**. The power supply reference point for each of these three devices is at a different potential, therefore requiring isolation.

The three power electronic switches connected to the negative terminal of the DC link **all** have their emitter, source or cathode terminals connected to the negative bus, and so a single power supply could be used for all three device driver circuits, hence the minimum of 4 isolated power supplies shown in the table. However, it is more common to use 6 identical power supplies to operate the device driver stages, as there are benefits in terms of modularity and commonality of wiring.

There are two main methods for deriving these device driver power supplies. The first is to provide the six isolated supplies from either a mains frequency transformer or a SMPS, in the same way all other control power is produced. An alternative is to provide a single high frequency square wave supply, which is coupled directly into the six driver circuits through dedicated high frequency (usually toroidal) transformers that are part of each driver circuit. Separate rectifier and regulation circuits then provide the necessary plus and minus supplies for each driver stage.

The cooling fans for the converter heat-sinks can be powered from the SMPS or directly from the mains, whichever is a cheaper solution. The major drawback of the mains supply is the inability to deal with the different mains voltages and frequencies which are found throughout the world. This can usually be solved by supplying the fan through an auxiliary transformer with several primary connections to match the most common voltage options.

6.3 The DC bus charging control system

A modern PWM-type AC drive operates with a fixed voltage DC bus. The fixed DC bus voltage is normally obtained from via a 6-pulse diode rectifier bridge from a 3-phase power supply. This voltage is usually 415 volts, 3-phase, 50 Hz while in some countries, the voltage is 380 volts, 3-phase, 50 Hz.

When the mains power is first connected to the input terminals of the AC drive, very high inrush currents would occur as the bank of filter capacitors across the DC bus charge. While the diodes in the rectifier module and the capacitors may be able to withstand these high currents, it is quite possible that upstream fuses or circuit breakers would operate to trip out the VSD. Therefore, some provision needs to be made to limit this inrush current. The *DC bus pre-charge circuit* is normally provided for this purpose.

There are two main approaches to solving the problem of inrush current:

- Pre-charge resistors, with a bypass contactor, either on AC side or DC side of the AC/DC rectifier bridge
- The AC/DC rectifier can be a controlled rectifier bridge instead of an uncontrolled diode bridge

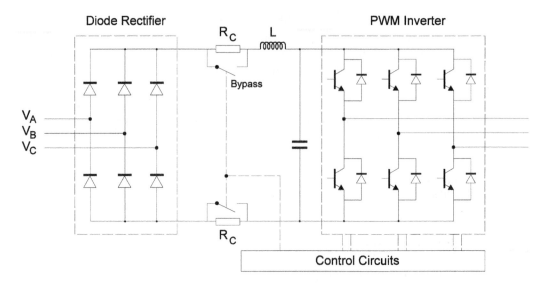

Figure 6.3:
Example of a DC bus pre-charging circuit

The first method is the most common method, an example of which is shown in Figure 6.3. Charging resistors are inserted between the input supply and the capacitor bank to limit the current when power is first applied. Once the capacitors are charged, these resistors would introduce additional losses in the VSD and therefore need to be bypassed during normal operation. A relay (small VSDs) or contactor (large VSDs) is used to bypass the charging resistors and carry the full rated current of the drive.

The control of the relay may be either via a simple timer circuit with a fixed time delay between power being applied and the inverter stage being enabled. A better method is to monitor the DC bus voltage and the bypass relay is closed after a certain voltage level has been attained. In the better quality VSDs, feedback may be provided from each of the power supplies in the central controller to verify their status.

Some form of interlock needs to be provided to ensure that the relay is closed before allowing the inverter stage to operate. If not, the high load current through the VSD will heat up and burn out the charging resistors. In addition, it is critical that all power supplies have had the opportunity to stabilize and establish regulation. As a result, most VSDs have a *start-up lock-out* circuit that delays starting for a short period after the VSD is powered up.

There are many variations on this theme, for example the resistors and relay can be either in the DC link or the 3-phase supply lines. There may be a single set of large resistors and one large relay or there may be multiple sets of smaller resistors and relays. Other variations of this technique include the use of semi-conductor bypass switches.

The main advantages of this method are:

- Simplicity of the control circuit
- Cheap and easy to implement

The main disadvantages of this method are:

- The losses associated with the relay contacts and coils
- The physical size of these components

- The reliability of these electromechanical devices, particularly when the motor control system requires a high number of energization and de-energizations

The second, less common, approach is to replace the normal diode rectifier with a phase-controlled rectifier bridge. This allows the capacitor voltage to be increased gradually, by controlling the firing angle, and thereby controlling the inrush current. This method is most often used on VSDs with larger power ratings above about 22 kW.

The main advantages of this method are:

- Conduction losses are lower
- Physical size is reduced by not having the relay

The main disadvantages of this method are:

- Power thyristors are more expensive than power diodes
- The control circuit is more complex in comparison with the relay circuit
- There is potential for false triggering of the phase control circuit due to notching and other disturbances on the mains
- Overall reactive power requirements are slightly higher

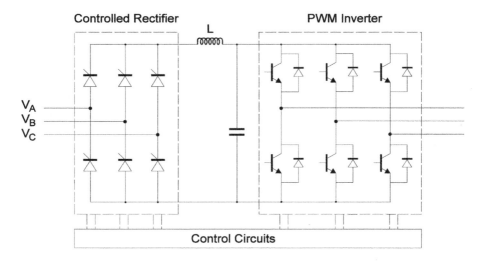

Figure 6.4:
DC bus charging using a phase-controlled thyristor bridge

6.4 The PWM rectifier for AC converters

A conventional AC VVVF converter is made up of 5 main sub-assemblies:

- **Ac/DC converter**, usually comprising a diode rectifier, for converting the 3-phase AC voltage to a DC voltage of constant amplitude. In some cases a phase-controlled thyristor bridge is used for DC bus charging. Once full DC voltage is achieved, the thyristor bridge is controlled to behave as a diode bridge.
- **The DC link**, usually comprising a DC choke, DC capacitor and a DC bus, for maintaining a smooth fixed DC voltage for the inverter stage.

- **The DC/AC inverter**, comprising a semiconductor bridge, for converting the DC voltage to a variable frequency variable voltage AC output.
- The **power supply modules**, for providing power to the control circuits for the interface system and the inverter switches.
- The **digital control system**, comprising the sequence control, internal control loops, protection circuits and user interfaces.

The **AC/DC rectifier** at the front end of the converter supplies the DC bus and capacitor with voltage from the AC mains supply. Using a 6-pulse diode bridge rectifier for this purpose has two main disadvantages:

- The AC line current waveform is non-sinusoidal (refer to Chapter 4) and is the source of odd harmonics, such as the 5th, 7th, 11th, 13th, etc. This high level of interference can couple to other equipment and disturb their normal operation.
- The harmonic current distortion results in a distortion of the voltage at the point of common coupling (PCC) which, if high enough (large VS drives), can affect the performance of other electrical equipment connected to the power supply system.
- Full four-quadrant operation is difficult with the diode rectifier because electrical power can only be transferred in one direction (refer to Chapter 3), which makes regenerative braking impractical with a standard AC VVVF drive.

If a PWM-type controlled rectifier (also called an 'active front end') were used, it could provide a solution to many of these problems. The 6-pulse PWM bridge converter with IGBTs is shown in Figure 6.5 and is electrically similar to a normal PWM inverter. As with the PWM inverter, it can transfer electrical energy in either direction, depending on the switching sequence of the IGBTs. For correct operation, it requires some minimum value of inductance in the line to avoid damage to the power semiconductor devices during switching. Line chokes may need to be added if the supply has a high fault level (low source impedance).

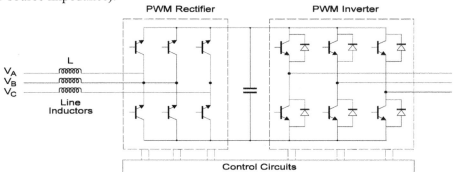

Figure 6.5:
The PWM rectifier for AC converters

One of the main problems of AC to DC power conversion using thyristor bridges, is the poor *displacement factor* due to phase-shifting of the current relative to the voltage and a bad *power factor* due to the distortion of the AC current waveform, which is non-sinusoidal. The PWM converter is capable of correcting both of these problems by drawing nearly sinusoidal current from the mains at unity displacement factor.

The unity *displacement factor* is achieved by forcing the current waveforms to exactly follow the fundamental voltage waveform at fundamental frequency, usually 50 Hz. The in-phase component of current is controlled to maintain the capacitor voltage at a required level, while the out-of-phase (quadrature) component of current can be made to be zero or leading to give a degree of power-factor correction for other loads.

Using pulse width modulation techniques, the current waveform can be made to be relatively undistorted (sinusoidal) and the reactive power requirement due to distortion will also be eliminated. This is assisted by the filtering effect of the line inductance.

6.5 Variable speed drive control loops

An AC frequency converter is designed to *control both the voltage and frequency fed to the motor* and is therefore often called a variable voltage variable frequency (VVVF) controller. The digital control system automates this process. For example, when an operator selects a speed setting on a potentiometer, the VSD control system implements this selection by adjusting the output frequency and voltage to ensure that the motor runs at the set speed. The accuracy of the control system and its response to the operator's command is determined by the type of control system used on that particular VSD.

The type of control used in VSD control systems follows an approach similar to that used in normal **industrial process control**. The level of control can be:

- **Simple open-loop control**, no feedback from the process
- **Closed-loop control**, feedback of a process variable
- **Cascade closed-loop control**, feedback from more than one variable

6.5.1 Open-loop control

The purpose of an electrical VSD is to convert the *electrical energy* of the mains power supply into the *mechanical energy* of a load at variable speed and torque. In many applications, VSDs are simply required to control the speed of the load, based on a setpoint command provided by an operator or a process controller.

Conventional VVVF converters are voltage source devices, which control the magnitude and frequency of the **output voltage**. The current that flows depends on the motor conditions and load, these are not controlled by the AC converter, but are the result of the application of voltage. The only current control that is exercised is to limit the current when its magnitude reaches a high level, for example at 150% of full load current.

There is no provision made for *feedback* of speed information from the motor to check if it is running at the required speed or if it is running at all. If the load torque changes, and slip increases or decreases, the converter would not adjust its output to compensate for these changes in the process.

This method of **open-loop control** is adequate for controlling steady-state conditions and simple applications, such as centrifugal pumps & fans or conveyors, which allow a lot of time for speed changes from one level to another and where the consequences of the changes in the process are not severe.

6.5.2 Closed-loop control

In industry, there are also those more difficult applications, where speed and/or torque must be continuously and accurately controlled. The required accuracy of the control is important and can have a large influence on the choice of drive technology. For those drive applications that require tight dynamic control, closed-loop control is necessary.

This type of performance can be achieved with closed loop vector control AC drives and standard DC drives.

Standard VVVF AC drives can be used in closed-loop control systems, such as pumping systems, which regulate pressure or flow, but in general these applications are not capable of high performance.

The typical configuration of a *closed-loop VSD system* is shown in Figure 6.6 and consists of the following main components:

- The **motor**, whose role is to convert the electrical energy of the supply into the mechanical energy necessary to affect the load
- A **transducer for measuring** the load quantity, which is to be controlled. This is used as a *feedback* signal to the control system. Where **speed** is important, the transducer can be a tachometer (analog system) or an encoder (digital system).

Where position is important, the transducer is a resolver (analog system) or an absolute encoder (digital system). However, there are less expensive means for measuring speed and position, depending on the required accuracy. Where current is important, the transducer is a current transformer.

- A **converter** which controls the flow of electric power to the motor. This is achieved with a *power electronic converter*, involving solid-state devices switching at high frequency under the control of a digital circuit.
- A **controller**, which compares the desired value of speed or position, called the *set-point* (SP), with the measured value, called the *process variable* (PV) and then gives a **control output** which adjusts the speed and torque to reduce the error (SP–PV) to zero. Previously, controllers were implemented by analog circuits, using operational amplifiers (Op-Amps). Modern controllers are implemented using microprocessors and digital circuits.

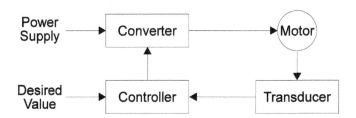

Figure 6.6:
Schematic of a closed-loop VSD control system

The desired value of the load, such as the speed, can be set manually by an operator turning a potentiometer (analog system) or by dialing up a value on a keypad (digital system). If the VSD is part of a complex control system, the desired value can be passed down from the *process control system (PLC or DCS)*, either by means of a 4–20 mA signal (analog system) or by means of a serial data link (digital system).

If each quantity in the control loop was directly proportional to the quantity before it, simple *open-loop control of speed* would be adequate, without the need for feedback of the process variable (PV). In AC drives, if accurate speed control is required, then feedback of the torque and speed variables is necessary. In particular, the motor current responds to an increase in motor frequency with a rise time dependent on its leakage inductance. On the other hand, the motor speed follows the torque with a rise time dependent on its inertia. While these inaccuracies may be acceptable in simple applications, such as pump speed control, it may not be acceptable for other difficult applications, such as the variable speed drives in a paper machine, where several drives operate in tandem. In these difficult applications, improved performance can be obtained with the use of several closed-loop control systems working together, known as multi-loop or cascade control.

This type of **closed-loop control system** has been redrawn in Figure 6.7 to emphasize the most important control aspects. The term *closed-loop feedback control* emphasizes the nature of the control system, where *feedback* is provided from the output back to the input of the controller.

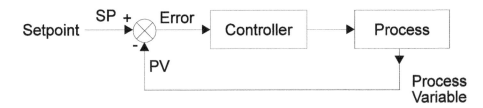

Figure 6.7:
A closed-loop feedback control system

In a closed-loop variable speed drive, the following takes place:

- Measurement of the *process variable* using an encoder
- Comparison of the *process variable* (measured speed) with the *set point* (desired speed) to give an error signal. *SP – PV = error signal*
- This *error signal* is then processed by the controller to adjust the output signal to the process, in this case, the AC converter, motor and speed transducer.

Figure 6.7 could be misleading because, in practice, the error point is usually part of the controller

6.5.3   Cascaded closed-loop control

For the difficult applications, which require very close speed and torque control, with a fast response to changes in the process, a single-loop controller may not be adequate to anticipate all the delays in the process. These make the controller difficult to design and difficult to setup during commissioning. Fortunately, a technique that deals with the problem in several smaller steps has evolved from past experience with DC drives. The solution consists of two *cascaded closed-loop controllers*. The basic setpoint is the speed setpoint, which is set by an operator via a potentiometer or from a PLC. Rather than attempting to calculate the desired inverter frequency directly to meet these speed requirements, a DC drive achieves this in two stages.

- The first **speed control loop** uses the speed error to calculate the desired torque setpoint to either increase speed (accelerate) or decrease speed (decelerate). The **speed control loop** only has to allow for one of the time delays in the system, which is the delay between the torque and the measured speed. This compensates for the mechanical transients in the system, mainly load inertia.

- The second **torque control loop** compares this set torque, the output of the speed controller, with the actual measured value and calculates the desired output frequency. The measured process variable in this case is the measured motor current, which is proportional to the motor torque. Therefore, this control loop is often called the *current loop*.

A vector control drive uses a similar strategy. In the design of the *torque control loop*, it is assumed that the rate of change of current is much faster than the rate of change of speed. This is equivalent to assuming that the motor is running at a constant speed. Consequently, the *current loop* only has to allow for the time delay between the output frequency and the current. As well as giving the desired inverter output frequency, it also gives the desired inverter voltage since the two are related. Both quantities are passed to the PWM switching logic, which controls the inverter switching sequence and speeds (see Figure 6.8).

The **current control loop** compensates for the electrical transients, mainly the winding inductance and resistance.

The block diagram of the cascaded loop controller comprises:

- An outer (major) speed control loop
- An inner (minor) torque control loop

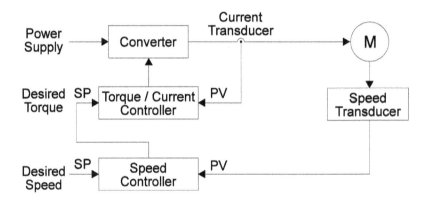

Figure 6.8:
Cascade controller for speed and torque

A major advantage of the *cascaded controller* is that it is possible to impose a current limit on the drive output by placing a limit on the input to the current/torque loop. This is usually set to prevent the speed control loop from asking for any more than about 150% of rated current from the current loop.

The current loop can respond quickly, in less than 10 ms. The speed loop responds more slowly because motor and load inertia are usually substantial. A response time of about 100 ms is typical for the speed control loop.

The response of the amplifier for the speed or current controller to a *step change* is shown in Figure 6.9. The simplest type of controller amplifier is one whose output is proportional to the input, called a proportional amplifier or P-control. P-control is not used in speed and current control loops because it does not respond well to the requirements for high accuracy and a fast dynamic response. It is more common in high performance VSDs to use proportional-integral control, or PI-control. The step response for PI-control consists of a combination of the step output of the P-control and a ramp due to the integral control.

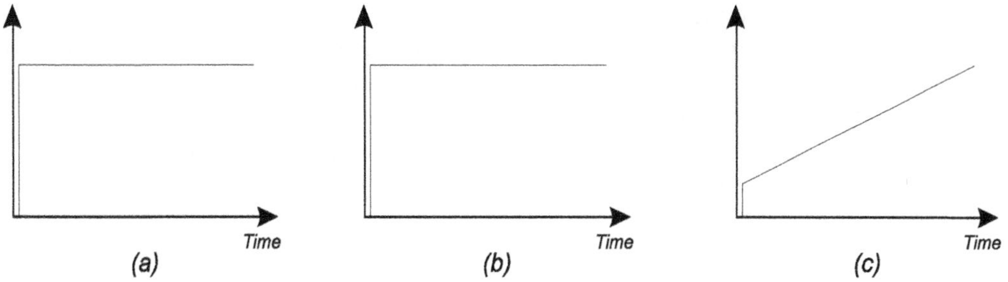

Figure 6.9:
Response of P and PI controllers to a step input

(a) Step input control change

(b) Proportional controller output

(c) Proportional/Integral controller output

6.6 Vector control for AC drives

The term *vector control* is probably one of the more abused terms in industrial control and consequently has caused considerable confusion amongst users of VS drives. *Vector control* for AC variable speed drives has been available from some drive manufacturers since the mid-1980s. The technique of *vector control* has only become possible as a result of the large strides made in solid-state electronics, both with microprocessors and power electronics.

It has been promoted as an AC drive equivalent to DC drives and claimed to be suitable for even the most demanding drive applications and this is where the confusion arises. The statement is true, but only to the extent that the principles of vector control are implemented. There are degrees to which this enhanced type of control can be applied to AC variable speed drives. Some manufacturers have encouraged this confusion in an effort to attribute higher performance characteristics to products that only partially apply the technology of *vector control*. The meaning of the various terms is covered later in this chapter after the fundamental principles of *vector control* are explained. Today, the term 'vector control' has become a generic name applied to all drives which provide a higher level of performance (compared to the fixed V/f drives).

Referring back to Chapter 2, electric motors produce torque as the result of the interaction of two magnetic fields, one in the fixed part (stator) and the other in the rotating part (rotor/armature) and their interaction across the air-gap. The magnetic fields are produced by the current flowing in the windings of the stator and rotor. The motor torque depends on the strength of both of these magnetic fields. In fact, torque is proportional to the product of the currents producing these two magnetic fields.

In a DC drive, it is fairly well understood that the output torque is proportional to the product of two current vectors, the armature current I_a (torque producing current) and the field current I_f (flux producing current), at $90°$ to one another. In practice, the field current (flux producing current) is normally held constant. Consequently, the armature current Ia is directly proportional to output torque of the motor. Armature current (I_a) can be used as the torque feedback in the cascaded closed-loop controller. Both these currents can readily be measured and accounts for the simple control of the DC drive.

In an AC induction motor (refer to equivalent circuit in Figures 2.5, 2.6 and 6.10), the flux producing current (I_m) and torque producing current (I_r) are 'inside' the motor and cannot be measured externally or controlled separately. As in the DC drive, these two currents are also roughly at 90° to one another and their vector sum makes up the **stator current**, which can be measured (Figure 6.11). This is what makes the vector control of an AC motor more difficult than its DC counterpart. The challenge for the AC flux-vector drive is to distinguish and control these two current vectors without the benefit of two separate circuits and only being able to measure and control the stator current.

The strategy of an AC vector control drive is to **calculate the individual current vectors** to eventually enable separate control of the flux current and/or the torque current under all speed & load conditions. As in the DC drive, the aim is to maintain a constant flux current in the motor.

The calculation of the current vectors involves the measurement of the available variables (such as the stator current (I_s), stator voltage (V_s), phase relationship, frequency, shaft speed, etc) and applying them to a 'motor model', which includes the motor constants (such as the stator resistance & inductance, the rotor resistance & inductance, the magnetizing inductance, number of poles, etc). Because of the many variables, there are many possible applications of a motor model, from simple estimation of motor conditions to those that are very comprehensive and very accurate. The more detailed the motor model, the more processing power is required.

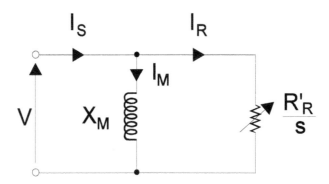

Figure 6.10:
Simplified equivalent circuit of an AC induction motor

Under **motor no-load conditions**, almost all the no-load stator current I_S comprises the magnetizing current. Any torque-producing current is only required to overcome the windage and friction losses in the motor. Slip is almost zero, stator current lags the voltage by 90°, so power factor is close to zero ($\text{Cos}\phi = 0$).

At **low motor loads**, the stator current I_S is the **vector sum** of the magnetizing current I_M (unchanged), with a slightly increased active torque-producing current. Stator current lags the voltage by a large angle ϕ, so power factor is poor ($\text{Cos}\phi \ll 1$). Slip is still small.

At **high motor loads**, the stator current I_S is the **vector sum** of the magnetizing current I_M (unchanged), with a greatly increased active torque-producing current, which increases in proportion to the increase in load torque. Stator current lags the voltage by the angle ϕ, so power factor has improved to be close to the full load power factor ($\text{Cos}\phi = 0.85$).

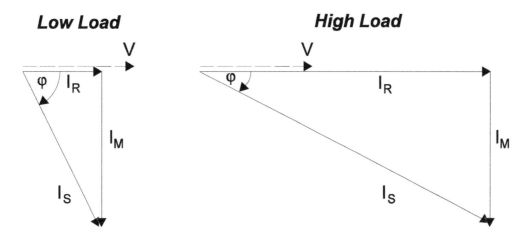

Figure 6.11:
Current vectors in an AC induction motor

Therefore, the central part of the vector control system is the *active motor model*, which continuously models the conditions inside the motor:

- Continuously calculates in real time the torque-producing current by implementing the following activities:
  - Stores the motor constants in memory to be used as part of the calculation
  - Measures stator current and voltage in each phase
  - Measures speed (with encoder) or calculates speed (no encoder)
- Continuously calculates in real time the flux-producing current
- Implements the speed control loop by comparing the speed feedback with the speed setpoint to provide an error output to the torque control loop
- Implements the torque control loop by comparing the active torque, calculated from the current and speed feedback, to provide an error output to the PWM switching logic controller
- Constantly updates this information and maintains tight control over the process.

For adequate dynamic response of the drive, the model calculations need to be done at least more than 2000 times per second, which gives an update time of less than 0.5 ms. Although this is easily achieved with modern high speed processors, the ability to continuously model the induction motor at this speed only became viable within the last 10 years or so with the development of 16-bit microprocessors. Initially, sufficient processing power for vector control was quite expensive, but over a period of time, the cost of the processors has come down and processing speed has increased significantly.

The main difference between a traditional fixed V/f ratio VVVF converter and a modern vector control drive is almost entirely in the control system and the extent to which the active motor model for vector control is implemented to control the switching pattern of the IGBTs of the inverter.

The power circuit for a vector converter is almost identical to that used by a VVVF drive:

- **Rectifier** to convert 3-phase AC to a DC voltage
- **Inductive choke** to reduce harmonics on the supply side
- **Dc link** with capacitor filter to provide a smooth and steady DC voltage
- An **IGBT semiconductor inverter bridge** to convert the DC to a PWM variable voltage variable frequency output suitable for an AC induction motor
- A **microprocessor based digital control circuit** to control the switching, provide protection and provide a user interface

Today, 'standard' AC variable speed drives from most reputable manufacturers implement vector control to some degree. For example, sensorless vector control is advertised as a performance feature with almost all modern AC drives.

There are essentially 3 basic types of control for AC variable speed drives today:

- Basic fixed V/f drive, provides fair speed control at a reasonable price and is suitable for the control of centrifugal pumps and fans
- V/f sensorless vector drive, provides better speed regulation, better starting torque and acceleration by implementing more/better control of the flux producing current vector (flux-vector)
- Closed loop field oriented vector control drive, provides excellent speed and torque control with DC like performance using cascaded PI control over speed, torque as well as flux regulation. Dynamic performance is excellent.

6.6.1   Basic fixed V/f drives

The control strategy of a fixed V/f drive is essentially **open-loop control** as shown below.

- The **speed reference** is taken from an external source and controls the voltage and frequency applied to the motor.
- The speed reference is first fed into a ramp circuit to convert a step change in the speed request to a slowly changing signal. This prevents electrical and mechanical shock to the speed control system. The acceleration and deceleration ramp times can be set by the user.
- The signal is then passed to a section that sets the magnitude of both the voltage and frequency fed to the motor. The V/f ratio between the voltage and frequency is kept constant at all times. It also sets the **rate of change** of these two values, which determines the motor acceleration.
- The base voltage and base frequency used for this ratio are taken from the motor nameplate.
- Finally, the signal passes to the PWM switching logic module, that controls the switching pattern of the IGBT switches to provide the voltage pattern at the output terminals according to the PWM algorithm (sine-coded, etc).
- There is usually no speed feedback from the motor. It is **assumed** that the motor is responding to and following the output frequency (open-loop control).
- The current feedback from the current transducer is there mainly for protection, indication and to set a current limit, it is NOT used as part of the control strategy.

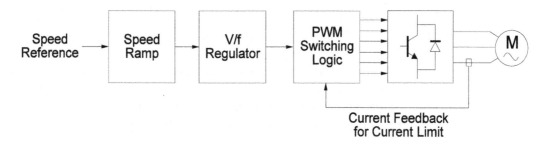

Figure 6.12:
Block control diagram of fixed V/f drive

It is necessary to monitor the stator current flowing to the motor. The drive usually monitors total current and cannot distinguish between I_m and I_r. This current is not used to control torque, but is aimed at the following functions:

- Measures actual current for the I^2t overload protection of the motor
- Provides protection of the power electronic components
- Provides a **current limit**, the control system reduces the frequency command signal when the current exceeds a predetermined value. Usually, current limit is set to 150% of the rated motor current.
- Some newer V/f drives provide **slip compensation** as a strategy for improving the speed holding capability in an attempt to maintain relatively constant motor speed even with changes in the motor load torque. As the output torque increases, the motor current increases, which can be used to adjust the output frequency of the converter. For example, at full rated load, the full slip value can be added to the output frequency. With *slip compensation*, improved speed regulation can be obtained from an induction motor without a speed feedback device.

This method of open loop fixed V/f control is adequate for controlling steady-state conditions and simple applications, such as pumps, fans and conveyors, which allow a lot of time for speed changes from one level to another and where the consequences of the changes in the process are not severe.

This type of drive is not well suited to the following:

- Applications where motors run at low speeds (below 5 Hz). The torque at low speed is generally poor because the stator volt drop significantly affects the magnitude of the flux-producing current. Many V/f drives include a 'start boost' when allows the V/f ratio to be boosted at starting in an attempt to improve the flux and consequently the starting torque.
- Applications which require higher dynamic performance.
- Applications that require direct control of motor torque rather than motor frequency.
- The dynamic performance of this type of drive with shock loads is poor.

6.6.2   V/f sensorless flux-vector drives (open loop vector)

The development of sensorless flux-vector drives was aimed at overcoming the main shortcomings of the fixed V/f drives, mainly the loss of torque at low speeds.

This type of drive is often also called an open loop vector drive because its basic core is still the fixed V/f ratio controller. But wrapped around this core are several additional control components:

- A **current resolver (mathematical model)** that uses the measured stator current to calculate (in real time) the two separate current vectors which represent the flux-producing current (I_m) and the torque-producing current (I_r)
- A **high performance current limiter** which uses the torque-producing current (I_r) to rapidly adjust the frequency command to limit current
- A **flux regulator** which continuously adjusts the V/f ratio to maintain an optimum control of the flux-producing current (I_m)
- A **slip estimator** that provides accurate estimation of the rotor speed based on the known motor parameters, without the use of an encoder. This provides improved slip compensation under all conditions of speed and load.

The result is greatly improved torque, particularly at low speeds, to provide high breakaway and acceleration torque and an improved dynamic response to shock loads. However, this type of drive does not provide torque control, it is still a speed control device. In addition, speed holding capability is substantially improved.

This type of drive can also be operated with an encoder, providing closed-loop control of the speed. This substantially improves the speed holding capability of the VS drive with speed regulation of 0.1%.

6.6.3   Closed-loop field oriented vector drives

Up to the end of the 1980s, high performance drive applications inevitably required the use of a DC drive. However, the high maintenance requirements of DC drives have encouraged the development of alternative solutions. Vector controlled AC drives have evolved to provide a level of dynamic performance that has now exceeded that of DC drives.

Closed-loop vector control is not required for every AC VSD application, in fact only on a minority of applications. But there are a number of applications that inherently require tight closed-loop control, with a speed regulation better than 0.01% and a dynamic response better than 50 radians/sec. This dynamic response is about 10 times better than that provided by standard V/f drives.

The control block diagram for a high performance vector control AC drive system is essentially a *cascaded closed-loop* type with speed and torque control loops:

- There are two separate control loops, one for **speed** and the second for **current**. This control strategy is similar to that used for the control of a DC drive.
  - **Speed loop** controls the output frequency, proportional to speed
  - **Torque loop** controls the motor in-phase current, proportional to torque
- The **speed reference** command from the user is first fed into a comparator, from where the error controls the speed regulator
- The speed error signal becomes the setpoint for the torque (current) regulator. This signal is compared to the calculated current feedback from the motor circuit and the error signal determines whether the motor is to be accelerated or decelerated

- There is a separate control loop for the flux current (V/f regulator)
- Finally, the signal passes to the PWM and switching logic section, that controls the IGBTs in such a way that the desired voltage and frequency are generated at the output according to the PWM algorithm (sine-coded, star modulation, VVC, etc).

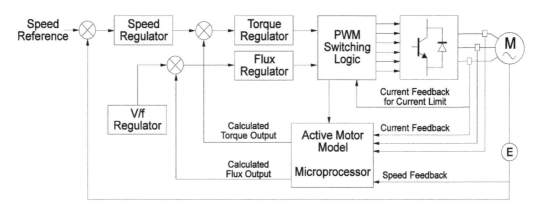

Figure 6.13:
Block diagram of the flux-vector converter control circuit

Although a shaft mounted incremental encoder can be used to measure speed in an AC drive, it is often considered to be an additional expense. In some cases it is difficult to mount on the motor, for example when motors have integral brakes. Even when an encoder is not used, the cascaded closed loop control can still be implemented because speed can be calculated by the active motor model, but with a lower level of accuracy due to the difficulty of calculating slip, particularly at very low speeds. Vector controlled drives which do not use encoders are usually referred to as **sensorless vector drives**. The dynamic response of vector control drives, which do not use an encoder, is usually inferior to those that do.

The following are some interesting figures that have been presented by one of the leading manufacturers of variable speed drives:

	DC drive with encoder	V/f flux vector sensorless	V/f flux vector with encoder	Field oriented sensorless	Field oriented with encoder
Speed accuracy	0.01%	1.0%	0.1%	0.5%	0.001%
Torque response	10–20 msec	100 msec	10–20 msec	1–10 msec	1–10 msec

Typical applications for this type of high performance VS drive are:

- Crane and hoist drives
- Rewinders on paper and steel-strip lines
- Paper machines

- Printing machines
- Positioning systems for automated manufacturing lines
- etc

When setting up high performance VSD controllers, a modest *proportional gain* gives a good transient response, while the *integral gain* gives high steady state accuracy. PI-controllers have the advantage that they can maintain a non-zero output to drive the converter although their input is zero. This is an advantage in closed-loop control because high accuracy should lead to zero error at the controller input.

Suitable values of *P* and *I* determine the step and ramp parts of the response respectively and have to be calculated for each inverter–motor–load combination.

- The **values of *P* and *I* for the speed loop** are dependent on the motor flux, load friction and inertia as they influence the response of speed to current.
- The **values of *P* and *I* for the current loop** depend on the inverter gain, motor resistance and leakage inductance, since they influence the response of current to the motor frequency.

In modern digital drives, the *P* and *I* values for both current and speed loops can be set by keypad or, alternatively, most modern digital drives usually include an algorithm for *self-tuning*. This removes the difficulties of 'tuning the loops', which was traditionally necessary with older analog DC drives. The *P* and *I* gains of the speed loop can be setup during commissioning to meet application requirements and seldom need to be changed.

There are a number of disadvantages of the vector controlled AC drive, when compared to a DC drive:

- The vector controller is far more complex and expensive when compared to the simple cascade controller of a DC drive.
- Encoder speed feedback is usually necessary to obtain accurate feedback of the motor shaft speed. Fitting these encoders to a standard squirrel cage AC induction motor is often difficult and makes the motor more expensive. In recent years, 'Sensorless' vector control has been developed where an encoder is not required. The approximate speed is calculated by the processor from the other available information, such as voltage and current. However, the speed accuracy and dynamic response of these drives is inferior to those using encoders.
- The nature of the drive itself often requires the AC motor to operate at high torque loadings at low speeds. The standard squirrel cage AC induction motor then requires a separately powered cooling fan, installed at the ND end of the motor.
- Regenerative braking is more difficult with a vector drive than with a DC drive. Resistive type **dynamic braking** systems are most often used with AC vector control drives.

6.7 Current feedback in AC variable speed drives

6.7.1 Methods of measuring current in variable speed drives

Current feedback is required in AC variable speed drives for a number of purposes:

- **Protection**, short circuit, earth fault and thermal overload in motor circuits

- **Metering,** for metering and indication for the process control system
- **Control**, current limit control and current loop control. Several methods have been developed over the years to measure the current and convert it into an electronic form suitable for the drive controller. The method chosen depends on the required accuracy of measurement and the cost of implementation. The main methods of measurement are as follows:
- **Current shunt**, where the current is passed through a link of pre-calibrated resistance. The voltage measured across the link is directly proportional to the current passing through it. This method was often used in drives with analog control circuits.
- **Hall effect sensor**, where the output is a DC voltage, which is directly proportional to the current flowing through the sensor. High accuracy and stability over a wide current and frequency range are amongst the main advantages of this device. This device is commonly used with modern digital control circuits.

The performance of a normal core type current transformer is usually not adequate for power electronic applications because its performance at low frequencies is poor and accuracy of measurement of non-sinusoidal waveforms is inadequate. The main methods of current measurement are described in detail in Appendix B.

6.7.2   Current feedback in general purpose VVVF drives

The primary need for current feedback in general purpose VSDs is inverter switching device protection. During short circuit or earth fault conditions, the device current will rise rapidly. If the power electronic switching device, such as an IGBT, BJT, GTO or MOSFET is not switched off quickly, it will be damaged and will fail. VSD reliability depends on the fast and accurate sensing of over-current conditions.

The secondary need for current feedback is to perform *current limiting*. Early versions of AC VVVF converters did not have a current limiting feature and would simply shut down if the load became too high, requiring manual reset by an operator. This increased downtime and gave VVVF converters a poor reputation in many industries, where overload trips were common. Modern VSDs use current feedback to limit the output current when high loads are encountered.

Current limiting is not the same as current control. Current control means that the current is being controlled at all times, whether it is high or low. Current limiting means that some action is taken to stop the current exceeding the desired limit point. This action may be only indirectly related to current, such as a change in frequency or voltage.

A third need for current feedback is to provide a current signal roughly proportional to load. This signal may be used internally by the drive to optimize motor volts/hertz or provide *slip compensation*, where the frequency is increased slightly as load increases to improve speed regulation. The signal may also be made available for external use, by the user, as a load indication signal. As outlined earlier in this chapter, the stator current of the motor is only roughly proportional to the mechanical load, since the stator current is the vector sum of the magnetizing current I_M and the torque-producing current I_R.

Motor current feedback can also be used to provide thermal protection of the motor. This requires a thermal model of the motor to be implemented in the drive control system, using frequency and current feedback and motor parameters to estimate the internal temperature of the motor, using an I^2t replica in the converter. If current level exceeds a set point for a period of time, the motor protection will trip the drive and give an indication of a motor thermal overload.

6.7.3   Current feedback in high performance vector drives

High performance drives, such as vector controlled drives, employ field oriented control and require current feedback as an integral part of their control loops. In these cases motor current is not simply limited at a pre-defined level. It is controlled to match a continuously changing torque demand. The vector components of the stator current in each phase are calculated, which requires current from all three phases. This can be achieved preferably with one hall effect CT in each output phase or alternatively two in the output phases and one on the DC bus. If only two-phase sensors are used, the third phase can be calculated from them, however the bus current sensor is still required for device protection.

High accuracy motor current feedback is also necessary to provide control of motor torque. Torque control is necessary in applications such as rewind/unwind systems, hoists, winches, elevators, positioning systems, etc.

6.7.4   DC bus current feedback

DC bus current feedback is suitable for switching device protection and current limiting in most AC VSDs. To a lesser extent, it can provide some load indication if suitably scaled. However, this is usually only accurate over a narrow range of speeds and loads, as the signal must be *synthesized* from the bus current waveform. It is the preferred method in general purpose drives, as it only requires a single current feedback device, reducing complexity and cost.

Robust performance for a large variety of load types can be achieved through careful implementation of DC bus current limiting. This is achieved by controlling the motor frequency to maintain the bus current at or below the preset limit point. For example, excessive loads may be encountered if a high inertia load is accelerated too quickly. This may occur if the acceleration time on the VSD is set without regard to the load dynamics.

For example, consider an application where a 22 kW motor would take 10 secs to accelerate a high inertia load at 150% rated torque and current. If the operator sets the acceleration time to 5 seconds, this would require 300% rated torque and around 500% current to accelerate the load. Clearly, a drive rated at 150% current overload will not be able to achieve the desired acceleration time. In this situation, a modern well designed VSD will not trip, but will modify its acceleration time to maintain the DC bus current at the current limit point. While the operator may not have been able to achieve the desired acceleration time, this is clearly preferable to the drive tripping on over current every time it starts.

6.8   Speed feedback from the motor

In closed-loop speed control of electric motors and positioning systems, the speed and position feedback from the rotating system is provided by transducers, which convert mechanical speed or position into an electrical quantity, compatible with the control system.

The following techniques are commonly used today:

- **Analog speed transducer**, such as a tachometer generator (tacho-generator), which converts rotational speed to an electrical voltage, which is proportional to the speed, and transferred to the control system over a pair of screened wires.

- **Digital speed transducer**, such as a rotary incremental encoder, which converts speed into a series of pulses, whose frequency is proportional to speed. The pulses are transferred to the control system over one or more pairs of screened wires.
- **Digital position transducer**, such as a rotary absolute encoder, which converts position into a bit code, whose value represents angular position. The code is transferred digitally to the control system over a screened parallel or serial communications link.

Analog speed transducers are increasingly being replaced by digital devices, which are more compatible with modern digital control systems.

The main methods of speed measurement are described in detail in Appendix C.

7

Selection of AC converters

7.1   Introduction

Although manufacturers' catalogues try to make it as easy as possible, there are many variables associated with the selection and rating of the optimum electric motor and AC converter for a *variable speed drive (VSD)* application. In many cases, it requires considerable experience to get the selection right. The reason why it is difficult is because there is always an engineering trade-off between the following:

- The need to build in a margin of safety into the selection procedure
- The need to keep the initial cost to a minimum, by selecting the optimum type and size of motor and converter for each application.

This chapter covers many of the principles for the correct selection procedure for *AC variable speed drives*, which use PWM-type variable voltage variable frequency (VVVF) converters to control the speed of standard AC squirrel cage induction motors.

The following checklist covers most of the factors that need to be considered:

- The nature of the application
- Maximum torque and power requirements and how these change with speed
- Starting torque requirements
- The speed range - minimum and maximum speed
- Acceleration & deceleration requirements (Is braking necessary?)
- Compatibility with the mains supply voltage
- Environmental conditions where the converter and motor are required to operate, ambient temperature, altitude, humidity, water, chemicals, dust, etc
- Ventilation and cooling for the converter and motor
- Direction (uni- or bi-directional)
- Accuracy of the speed control
- Dynamic response (speed and torque response requirements)

- Speed regulation requirements with changes in load, temperature, supply voltage
- The duty cycle, including the number of starts and stops per hour
- Overall power factor of the drive system and its effect on the mains supply
- EMI and harmonics in the mains power supply and in the motor and motor cable
- Are EMI filters required?
- Earthing, shielding and surge protection requirements
- Torque pulsations in the rotor shaft
- Control method - manual, automatic, analog, digital, communications
- Control and communications interfaces required for the plant control system
- Indications required
- Reliability requirements, is a dedicated standby unit required
- Protection features, in-built and external features required
- Power and control cable requirements
- Parameter settings, local or remote programming
- Maintenance, spares and repair considerations
- Cost of the alternative systems, taking into consideration the capital cost, performance advantages, energy savings, efficiency or process improvements.
- Noise due to the harmonics in the motor
- Mechanical resonance at certain motor speeds

This chapter covers many of the technical issues that need to be considered, but will not be able to address all the above factors in detail.

7.2   The basic selection procedure

Experience has shown that most of the problems experienced with AC VSD applications can usually be attributed to human error, mainly.

- The incorrect selection and rating of the AC induction motor
- The incorrect selection and rating of the AC converter
- The incorrect parameter settings installed in the VSD control system

As with all other electrical equipment, it is essential that the drive be correctly rated to do the job under all anticipated circumstances. The AC variable speed drive system is correctly selected and rated when:

- Motor specification is correct

The correct type and size of electric motor has been selected, whose output torque, speed and accuracy are adequate for all load and environmental conditions.

- Ac converter specification is correct

The correct type and size of AC converter has been selected, whose output (voltage, current, frequency) meets the motor requirements for all load and environmental conditions.

Usually, too much emphasis is placed on the selection of the converter, which is the expensive part, while too little emphasis is placed on the selection of the motor.

The correct procedure is:

- The first step is to select a correctly rated electric motor
- Only when this is completed, a suitable AC converter is chosen to match the requirements of the motor

From the motor point of view, the main factors which need to be considered are the motor power rating (kW), the number of poles (speed) and the frame size so that the load torque on the motor shaft remains within the continuous torque capability of the motor at all speeds within the speed range. High torques of short duration, such as starting torque, can usually be easily accommodated within certain limits outlined below.

7.3 The loadability of converter fed squirrel cage motors

When selecting an AC motor for any drive application, the most important requirement is to ensure that the motor does not become overloaded or stall under all circumstances of speed and load, i.e. over the entire speed range.

To stay within the temperature rise limits of the motor, the torque required by the load for starting, acceleration and for continuous running must be within the rated output torque capacity of the motor.

For AC motors connected to the power supply direct-on-line (DOL), it is usually sufficient to ensure that load torque is sufficiently below motor torque at the rated speed of the motor, for example the torque at 1450 rev/m on a 4 pole motor. These fixed speed drives operate only at one speed. It may also be necessary to ensure that the starting torque of the motor is higher than the breakaway torque of the load.

In the case of a variable speed drive, the load torque usually changes with speed, so it is essential to check that the motor torque exceeds the load torque at all speeds in the speed range. For example, a centrifugal pump has a variable torque characteristic, where the starting torque is low and the torque increases as the square of the speed as shown in Figure 7.6. Other loads, such as a conveyor, may have a constant torque characteristic, where the load torque remains constant for all speeds, as shown in Figure 7.7.

The continuous load torque capacity (loadability) of a standard TEFC squirrel cage induction motor used with VVVF converters is always lower than the rated torque of the motor itself for the following reasons:

- **At all speeds**, the load capacity is reduced as a result of additional heating in the motor caused by harmonic currents, however small. These occur because the output current waveform of the converter is not completely sinusoidal, even with modern PWM inverters with switching frequencies around 10 kHz.

  Traditionally, a de-rating of between 5% and 10% was used, depending on the type of motor (number of poles) and the type of converter. But, it has become common practice with modern PWM inverters to make provision for no de-rating at all. This relies on the fact that modern IEC motors always have a built-in thermal reserve (refer to Chapter 2), which will accommodate any additional heating. Also, the mechanical load is seldom exactly equal to the motor rating and is often lower by as much as 20%.

  It is considered good engineering practice to allow a small *margin of safety*, so a de-rating of up to 5% is usually provided.

Consequently, the overall output torque of the VSD, running at its **base speed** of 50 Hz, is taken to be about 95% of the motor catalogue's rated torque at 50 Hz and at rated DOL supply voltage.

- **At speeds below base speed**, in the speed range 0–50 Hz, the motor's continuous load capacity is reduced because of decreased fan cooling of both the stator and the rotor. The reduction in continuous torque output depends on the type and size of the motor, but in the absence of other de-rating tables, can be assumed to reduce to about 40% of rated torque at standstill. Some natural radiated and convectional cooling from the motor frame takes place even when stopped.

  For some constant torque applications, a separately powered auxiliary cooling fan mounted onto the motor can be used to improve stator cooling and increase load capacity at low speeds. These are usually designed to provide a motor with a volume of air equal to that flowing at rated speed for a motor of equal frame size. This supplementary cooling does not entirely overcome the load capacity problem. In a squirrel cage motor, the rotor losses are usually higher than the stator losses and the rotor losses become difficult to dissipate at low speeds even with supplementary cooling. Separate cooling is more effective with open motors (IC01).

- **At speeds above base speed**, the output torque capability of the motor is reduced because of reduced air-gap flux (lower magnetic field). The output torque reduces in direct proportion with the motor speed above 50 Hz. (Refer to next Section 7.4.)

The AC VSD **loadability curve**, shown in Figure 7.1 below, summarizes the factors above and the solid line marks the maximum limits of continuous load torque. Motors fed from VVVF converters can be loaded continuously at torques below the loadability limit line for the speed range.

However, motors can tolerate load torques greater than the level permitted by the loadability curve for short periods of time. High torques are usually required during starting and acceleration up to the preset speed range. The duration of the allowed overload depends on several factors such as the size of the motor, the magnitude of the overload and the speed. Many AC converters have an over-current capacity of up to 150% for 60 secs to cover starting and transient operation.

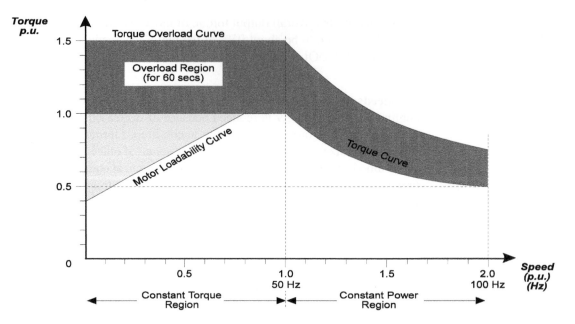

Figure 7.1:
The speed range and load torque capacity (loadability) of a TEFC squirrel cage motor when controlled by a PWM-type VVVF converter

The above curve shows the thermal load capacity of an AC VSD that is typical of curves used by many drives specialists. They are based on standard IEC-type squirrel cage motors running with PWM-type VVVF converters. The curves are given in per unit values, so they can be applied to motors of any voltage and size. Small motors below 5.5 kW have a slightly higher load capacity at low speeds.

The equivalent load power (kW) capacity curve is shown in the figure below. In the region below base speed, known as the **constant torque region**, the power capability increases linearly from zero at zero speed to full power at the base speed.

Above base speed, the power output capability cannot increase further and remains constant for further increases in speed. This is known as the **constant power region**.

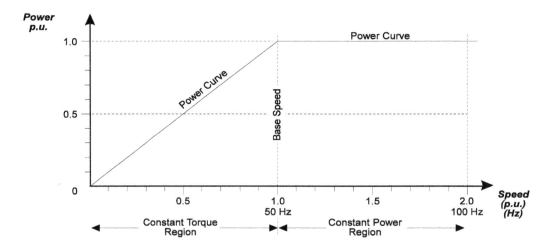

Figure 7.2:
Load power capacity of a TEFC squirrel cage motor when controlled by a PWM-type VVVF converter

7.4 Operation in the constant power region

As with the DC drive, the motor output torque of an AC VSD is proportional to the product of the air-gap flux and the torque-producing rotor current. The **stator current** can be taken to be roughly the same as the rotor current (refer to Chapter 6).

The control system of a typical *open-loop* VVVF drive system generates an output voltage with a constant V/f ratio, to provide approximately constant motor air-gap flux in the region between zero speed and the base speed (50 Hz). This produces a constant torque characteristic in the region between zero and the base speed of 50 Hz and the power increases in proportion to speed.

It is possible to increase the inverter output frequency beyond the base speed, in fact to frequencies as high as 400 Hz, in some converters. At speeds above the base speed, the output voltage remains constant at the maximum level possible from the fixed DC bus.

Consequently, the V/f ratio (air-gap flux) will fall in inverse proportion to the inverter frequency and the output torque of the motor falls in proportion with the flux. In this region, although torque is reduced, the output power remains constant and this region is known as the *constant power region* or *the field weakening region*. The latter name comes from the DC drive terminology, where speed can be increased above the base speed by reducing (weakening) the field flux. Essentially the same thing is happening here.

The main effect of operation above the base speed is the reduction in output torque of the motor in direct proportion to the increase in speed. In this region, precautions need to be taken to ensure that the motor torque does not fall below that of the load, otherwise the motor will stall. The figure below shows the torque–speed curve (dashed line) and power–speed curve (solid line) for the two regions:

- Constant torque region, below base speed
- Constant power region, above base speed (field weakening region)

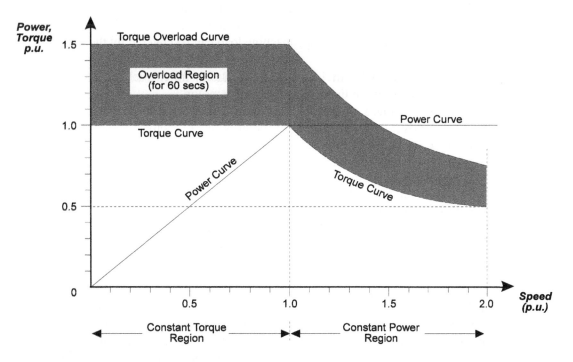

Figure 7.3:
Torque–speed and power–speed curves for an AC VSD

7.5   The nature of the machine load

No electric motor drive, fixed or variable speed, can be correctly specified without knowing something about the machine that is to be driven, specifically the **machine load**.

For fixed speed drives, it is often thought to be sufficient to specify only the *power requirement in kW* at the rated speed. On larger drives, motor manufacturers usually ask for more information about the load, such as the moment of inertia, to ensure that the design of the motor can cope with the acceleration requirements. In the case of AC variable speed drives, more details about the load characteristics are always necessary.

The output torque of an AC VSD is considered to be adequate when it:

- Exceeds the breakaway torque of the machine load
- Can accelerate the load from standstill to its preset speed within the acceleration time required by the process
- Exceeds the load torque by an adequate margin during continuous operation at any speed in the speed range and under all conditions
- Motor current does not exceed the thermal ratings of all electrical components and remains below the loadability curve during continuous operation.

The selection procedure outlined below applies mainly to single motor AC variable speed drives without special requirements and where the drive is continuous after an initial acceleration period. This means that a *standard* TEFC squirrel cage induction motor and a *standard* AC VVVF converter can be used. Multi-motor VSDs and other special applications require further investigation and will be considered in later sections. In general, special applications should be referred to the manufacturers of motors and/or AC converters for special rating calculations.

There are a great variety of different types of machine loads that are commonly driven by VSDs, each with different characteristics of torque, inertia, etc. Examples are pumps, fans, crushers, compressors, conveyers, agitators, etc. For VSD applications, most of what needs to be known about the machine load can be covered by the following:

- The **load torque**, the type, magnitude and characteristics of the load torque connected to the output shaft of the motor
- The **speed range**, the minimum and maximum speed of the variable speed drive
- The **inertia** of the motor and mechanical load connected to shaft of the motor

7.5.1   The load torque

The torque required by the driven machine determines the size of the motor because the continuous rated torque of the motor must always be larger than the torque required by the driven machine.

The magnitude of the load torque determines the cost of the motor because, as a rule of thumb, the cost of an electric motor is approximately proportional to its rated output torque (not its rated power!). The load torque is not necessarily a fixed value. It can vary with respect to speed, position, angle and time as shown in the table of Figure 7.4.

Another important aspect of the load torque is that the figure should apply at the shaft of the motor. When gearboxes, conveyers or hoists are involved, the actual torque at the machine must be converted to torque at the motor shaft. The conversion formulae are given in Figure 7.5 to convert the load torque, speed and moment of inertia to motor shaft values.

Machine Load	Characteristic Curve	Formulae
Conveyors screw Conveyors Pos. displ. pumps Compressors Ball mills	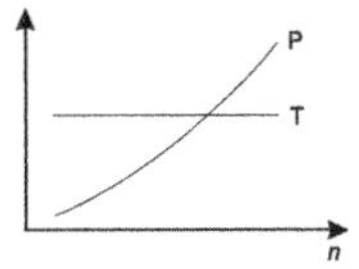	$T = k \quad (Constant)$ $P = k \cdot n \cdot T$
Centrifugal pumps Centrifugal fans	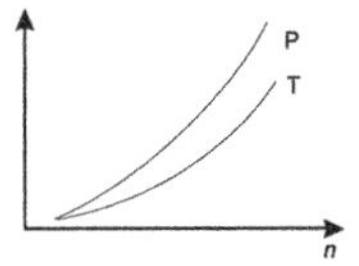	$T = k \times n^2$ $P = k \times n^3$
Extruders Slurry pumps	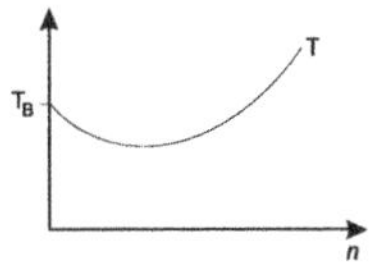	$T_B = Breakaway$
Winders Lathes	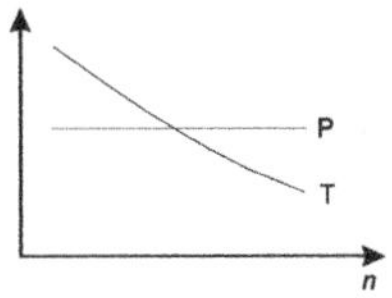	$P = k$ $T = \dfrac{k \cdot P}{n}$
Reciprocating Machines	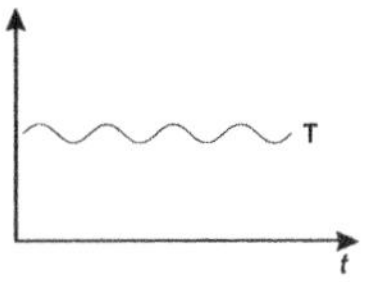	
Presses	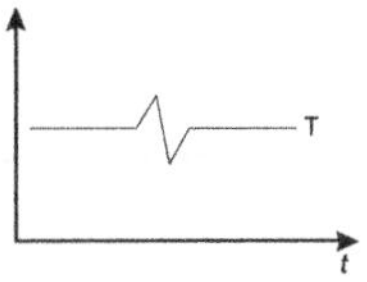	
Crushers Mills Wood chippers	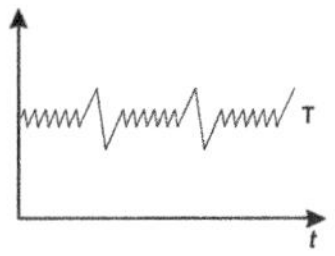	
Cranes Sawmills Presses	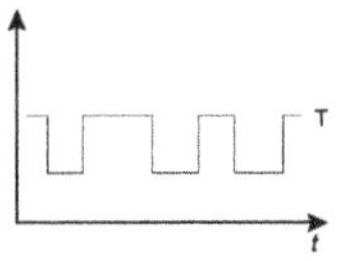	

Figure 7.4:
Torque characteristics of typical types of machine loads as a function of speed angle and time. (Note: k = constant)

Mechanical Device	Characteristic	Conversion Formulae
Gearbox	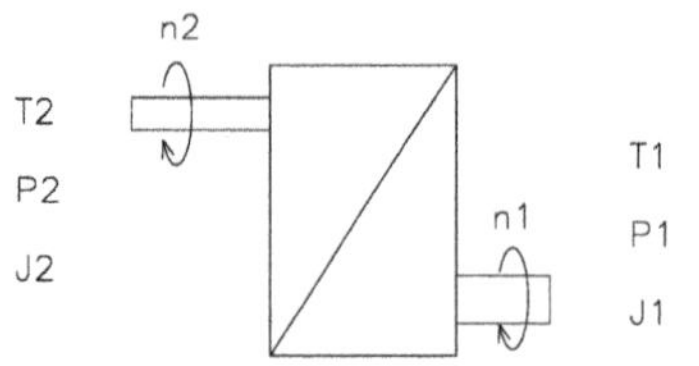	$T_2 = \dfrac{T_1}{\eta}\dfrac{n_1}{n_2}$ $P_2 = \dfrac{P_1}{\eta}$ $P_2 = \dfrac{P_1}{\eta}$
Conveyor	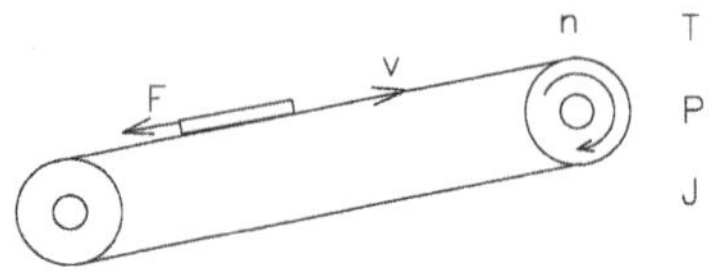	$T = \dfrac{F}{\eta}\dfrac{v}{2\pi n}$ $P = \dfrac{F\,v}{\eta}$ $J = M\dfrac{v^2}{(2\pi n)^2}$
Hoist	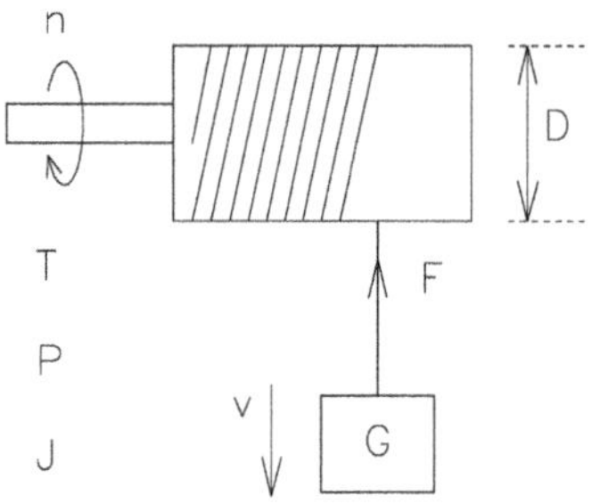	$T = \dfrac{F}{\eta}\dfrac{D}{2}$ $P = \dfrac{F\,v}{\eta}$ $J = M\dfrac{D^2}{4}$ $v = \pi\,D\,n$

T = Torque in *Nm* F = Force in *N*
P = Power in *kW* v = Velocity in *m/sec*
J = Inertia in *kgm²* η = Efficiency in *p.u.*

Figure 7.5:
Formulae to convert load torque, power and moment of inertia to motor shaft values

The load requirements are often given as the mechanical absorbed power (P_M kW) at a particular speed (*n* rev/m). The mechanical load torque may then be calculated from the following formula given in Chapter 2:

$$T_M = \frac{9550 \times P_M \text{ (kW)}}{n \text{ (rev/m)}} \quad \text{Nm}$$

Where T_M = Mechanical torque at the motor shaft in Nm
             P_M = Absorbed load power at the motor shaft in kW
             n = Actual rotational speed of the motor shaft in rev/min

7.5.2   Variable torque machine loads

Variable torque machine loads are those which exhibit a variable torque over their entire speed range, such as centrifugal pumps and fans. The torque–speed curve for these loads are shown in Figure 7.6 below.

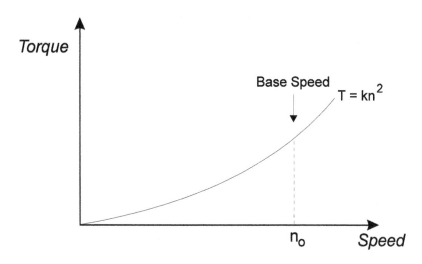

Figure 7.6:
Torque–speed characteristic of a variable torque load

The use of variable speed drives for the speed control of pumps and fans are the simplest applications and provide the least number of problems. The reason is that the breakaway starting torque is usually very low and then rises with speed.

The following are some of the important factors associated with this type of load:

- The starting torque for normal centrifugal pumps and fans is very low and below the loadability curve of the AC motor for all speeds. Slurry pumps can sometimes be a problem, as they can have a high breakaway torque.
- The required starting current is low, so the overload capacity of converters is seldom required during acceleration.
- Running for long periods at low speeds is seldom a problem.
- However, running at speeds above the motor base speed could be a problem, because the power requirement of this drive increases as the cube of the speed. This is incompatible with the capabilities of the constant power region.

The manufacturers of modern PWM converters have tried to reduce the cost of VSDs for pump and fan applications by providing reduced performance drives with the following features:

- Low over-current capability, typically up to 120% for 30 sec

7.5.3   Constant torque machine loads

Constant torque machine loads are those which exhibit a constant torque over their entire speed range, such as conveyors, positive displacement pumps, etc. The torque and power curves for these loads are shown in Figure 7.7 below.

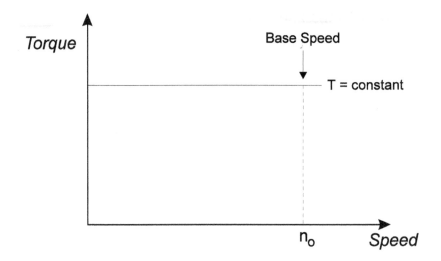

Figure 7.7:
Torque–speed characteristic of a constant torque load

The following are the potential problems when driving constant torque loads from a converter fed electric motor:

- The starting torque is theoretically equal to the full speed load torque but, in practice, the real starting torque can be much higher due to the additional requirements of:
  - breakaway torque
  - acceleration torque (dynamic torque)
- Running for long periods at low speeds can result in motor thermal over-load, if the load torque is above the motor loadability curve. Separate forced cooling may be necessary in some cases.
- Running at speeds above the motor base speed could also be a problem, with increased motor slip and a higher possibility of stalling the motor.

The manufacturers of modern PWM converters have tried to overcome some of these problems by providing the following features:

- High short time over-current capability, typically up to 150% for 60 sec, which is often required during starting
- Voltage boost to compensate for stator volt-drop at low frequencies
- Providing adequate motor protection to protect the motor from overload
  - Motor thermal protection which models low speed cooling reduction
  - Motor thermistor protection inputs

Some of these limitations are illustrated in the *loadability curve* of Figure 7.1.

7.5.4   The speed range

The selection of the correct size of electric motor for a VSD is affected by the speed range within which it is expected to run continuously. The important factor is that the motor should be able to drive the load continuously at any speed within the speed range without stalling or overheating the motor, i.e. the torque and thermal capacity of the motor must be adequate for all speeds in the speed range, within the loadability limits.

Running the motor at *below base speeds (f* < 50 Hz*)* with a standard TEFC cage motor has the following effects on the motor:

- Reduces the motor cooling because the cooling fan, which is attached to the motor shaft, runs at reduced speed. Therefore, the temperature rise in the motor will tend to be much higher than expected. (Refer to Figure 7.1.)

Figure 7.8 shows an example of the torque–speed curve for a variable speed pump drive, operating in the range from 10 Hz to 50 Hz. Some comments are:

- The load torque is well within the loadability limits at all speeds.
- The maximum speed is below the base speed of 50 Hz. The speed range should NOT be increased above 50 Hz because the load torque will exceed the loadability limit of the drive. (Load torque increases as the square of the speed.)
- Starting torque is low, so there should be no problems with breakaway.
- The acceleration torque is high, so the drive can be expected to quickly reach its maximum speed, if fast acceleration is required. However, with pumps, a long acceleration time is normally desirable to prevent *water hammer*.

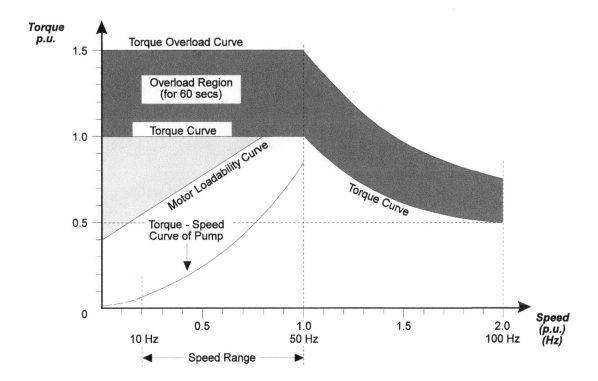

Figure 7.8:
Example of speed range and torque curve of a variable speed pump drive when controlled by a PWM-type VVVF converter

Running the motor at *above base speeds (f > 50 Hz)* with a standard TEFC cage motor has the following effect on the motor:

- The air-gap flux is reduced because the V/f ratio is reduced. Consequently, there is a reduction in the output torque capability of the motor. The torque is reduced in proportion to the frequency. The load torque is not permitted to exceed the pullout torque of the motor, even for a short period, otherwise the motor will stall.

The maximum torque allowed at above-synchronous speeds depends on the motor characteristics and frequency as follows:

$$T_{Max} \leq 0.6\, T_P \frac{50}{f} \quad Nm$$

where T_p = Pull-out torque (maximum torque) of the motor in Nm

f = Actual frequency in the above-synchronous range in Hz

0.6 = Factor of safety

Figure 7.9 shows an example of the torque-speed curve for a variable speed conveyor drive, operating in a similar range from 10 Hz to 50 Hz.

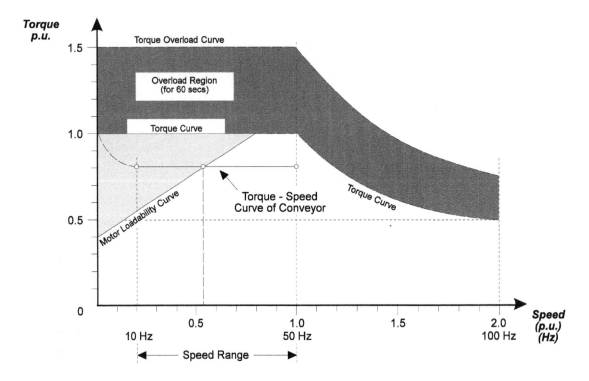

Figure 7.9:
Example of speed range and torque curve of a variable speed conveyor drive when controlled by a PWM-type VVVF converter

Some comments on this application are:

- The load torque falls outside the loadability limits at low speeds below 28 Hz. There could be problems running the motor continuously at speeds below 28 Hz.
- Although the maximum speed is below the base speed of 50 Hz, but the speed range could be increased above 50 Hz to take advantage of the loadability characteristic above 50 Hz. (Load torque remains constant with increases in speed.)
- Starting torque is high, with a high breakaway, so there may be some problems with breakaway.
- Acceleration torque is small, so the drive ramp-up time may have to take place over a long period to avoid exceeding the VSD current limit.

7.5.5 The inertia of the machine load

During acceleration and deceleration, the moment of inertia of the load imposes an additional *dynamic acceleration torque* on the motor. The dynamic acceleration torque is the extra torque that is required to change the kinetic state of the load as it accelerates from one speed to another. The moment of inertia and the required acceleration time together affect the motor torque and consequently the size and cost of the motor.

The dynamic acceleration torque T_A is calculated as follows:

$$T_A = J\, 2\pi \frac{dn}{dt} \quad \text{Nm}$$

where dn = Change in speed during acceleration in rev/sec
dt = The time it takes to effect the speed change sec
J = Moment of inertia of the drive system in kgm^2

From Chapter 2, this can be rewritten as follows, with the speed in rev/min:

$$T_A = J_{Tot} \frac{2\pi}{60} \frac{(n_2 - n_1)}{t} \quad \text{Nm}$$

When running at a constant speed, a motor must deliver a torque corresponding to the machine load torque T_L. During acceleration, there is the added requirement for the acceleration torque T_A. So the total torque required from the motor must be greater than the sum of the load torque T_L and the dynamic torque T_A. The motor must be selected to provide this total torque without exceeding its load capacity.

$$T_M \geq T_L + T_A \quad \text{Nm}$$

Example:

A conveyor drive is to be accelerated from zero to a speed of 1500 rev/min in 10 secs. The moment of inertia of the load J_L = 4.0 kgm^2. The torque of the conveyor load, referred to the motor shaft, is a constant at 520 Nm. The motor being considered is a 110 kW, 1480 rev/m motor with a J_M = 1.3 kgm^2. Is this motor adequate for this duty?

The total moment of inertia of the drive system is

$$J_{\text{Tot}} = 4.0 + 1.3 = 5.3 \quad \text{kgm}^2$$

During acceleration, the dynamic torque required is

$$T_{\text{A}} = 5.3 \frac{2\pi}{60} \frac{(1500 - 0)}{10} \quad \text{Nm}$$

$$T_{\text{A}} = 83.25 \quad \text{Nm}$$

The machine load is a constant torque type with a value given above as

$$T_{\text{L}} = 520 \quad \text{Nm}$$

During acceleration, the motor must supply a total torque T_{Tot} of

$$T_{\text{Tot}} = T_{\text{L}} + T_{\text{A}} \quad \text{Nm}$$

$$T_{\text{Tot}} = 520 + 83.25 = 603.25 \quad \text{Nm}$$

The rated motor torque may be obtained from the manufacturer's tables or calculated from the rated power as follows:

$$T_{\text{N}} = \frac{9550 \times 110}{1480} \quad \text{Nm}$$

$$T_{\text{N}} = 709.8 \quad \text{Nm}$$

Because $T_{\text{N}} \geq T_{\text{Tot}}$, the motor is evidently suited for the drive requirements.

When the motor drives the mechanical load through a gearbox or pulleys, the inertia of the load must be '*referred*' to the motor shaft using the formula given in the table in Figure 7.5.

$$J_M = J_L \frac{(Load\ Speed\ )^2}{(Motor\ Speed\ )^2} \quad \text{kgm}^2$$

Where J_M = Inertia at the motor shaft

J_L = Inertia at the load shaft

Example:

A 5.5 kW motor of rated speed 1430 rev/min and rotor inertia of 0.03 kgm^2 drives a machine at 715 rev/min via a 2:1 pulley and belt drive. The inertia of the mechanical load is 5.4 kgm^2, running at 715 rev/min at full rated speed. If the load is a constant torque load with an absorbed power of 4.5 kW at 715 rev/min, what is the acceleration time for this drive system from standstill to full load speed of 715 rev/min? Assume that the full motor torque is 150% of rated torque and is constant over the acceleration period.

The rated output torque of the motor is given by:

$$T_N = \frac{9550 \times 5.5}{1430} \quad Nm$$

$$T_N = 36.7 \quad Nm$$

The maximum output torque is 150% during the acceleration period

$$T_M = 1.5 \times 36.7 = 55.05 \quad Nm$$

The absorbed power of the load is 4.5 kW at 715 rev/m, which gives a load torque of

$$T_N = \frac{9550 \times 4.5}{715} \quad Nm$$

$$T_L = 60.1 \quad Nm$$

This needs to be converted to the motor shaft by the pulley ratio

$$T_{Lm} = 60.1 \frac{715}{1430} = 30.05 \quad Nm$$

The acceleration torque is the difference between maximum motor torque and load torque referred to the motor shaft

$$T_A = (T_M - T_{Lm}) \quad Nm$$

$$T_A = (55.05 - 30.05) = 25 \quad Nm$$

The Inertia of the mechanical load referred to motor shaft is

$$J_M = 5.4 \frac{(715)^2}{(1430)^2} \quad kgm^2$$

$$J_M = 1.35 \, kgm^2$$

$$J_{Tot} = 1.35 + 0.03 = 1.38 \quad kgm^2$$

If a gearbox is used, its inertia of the gearbox itself, referred to the motor, should also be taken into consideration.

Therefore, to calculate the overall acceleration time of the drive system, the simple formula above may be applied, provided that the acceleration torque remains constant and the drive accelerates linearly in a uniform time.

$$t = J_{Tot} \frac{2\pi}{60} \frac{(n_2 - n_1)}{T_A} \quad \text{sec}$$

Where t = Total acceleration time in sec

J_{Tot} = Moment of inertia of the (motor + load) in kgm^2

n = Final speed of the drive in rev/min

T_A = Acceleration torque of the drive system in Nm

Assuming that the acceleration torque remains constant over the acceleration period, the minimum acceleration time of the conveyor drive system is:

$$t = 1.38 \frac{2\pi}{60} \frac{(1430 - 0)}{25} \quad \text{sec}$$

$$t = 8.3 \, \text{sec}$$

This formula can be used as a rough estimate for acceleration time, but it is only an approximation because the acceleration torque is seldom a constant value, it is the difference in two changing values being the *motor torque* and the *load torque*.

There are two alternative methods of achieving a more accurate result:

- Use a computer program to accurately calculate the result. This technique is used by large engineering companies, motor manufacturers and vendors.
- Use a manual graphical system to calculate the acceleration time.

The first step in calculating the acceleration torque is to clearly define the motor and the load torque–speed characteristics.

The motor torque–speed curve is usually available from the manufacturer. If this is not available, important points on the curve are usually given in the motor manufacturer's catalogue in the form of the 4 points given below.

- Starting breakaway torque
- Pull-up torque and speed
- Pull-out torque (or breakdown torque) and speed
- Rated full-load torque and speed

7.6   The requirements for starting

Variable torque loads, such as centrifugal pumps and fans have a very low starting torque requirement and are easily pulled away and accelerated to the set speed by any VSD. The main area for concern is the **high breakaway torque** sometimes required on some pumps, such as slurry pumps, where some sediment can settle inside the pump during periods when the pump is stopped. The other limiting factor is the total absorbed power at full rated speed, which must be within the capacity of the drive.

Constant torque loads, such as conveyors and positive displacement pumps, are slightly more difficult because they require full torque at starting, but this does not usually present a problem. However, for some types of load, such as wood-chip screw conveyors, an additional breakaway torque may also be required to pull them away from standstill. Other examples of this are extruder drives and positive displacement pumps, particularly when used with congealing fluids. This high torque is usually of a temporary nature but the drive must be selected to ensure that the VSD can provide the necessary breakaway torque without stalling.

There are two main factors that affect the starting and low speed torque capability of a squirrel cage motor controlled by an AC VVVF converter.

- To avoid over-fluxing the motor, the V/f ratio must be kept constant. At low frequencies, the voltage applied to the stator of the motor is low to keep this V/f ratio constant. Referring to the equivalent circuit of an induction motor (Chapter 2), there is a volt drop in the stator winding and the air-gap flux is then significantly reduced. This affects the output torque of the drive. The problem can be relatively easily overcome by boosting the voltage at low speeds to compensate for the stator volt drop. Most modern converters provide a *torque boost* setting that may be adjusted by the user.

- Most VVVF converters have a current limiting control feature to protect the power electronic components against over-currents. So the maximum motor current is limited to the current limit setting on the converter. Since the motor torque is roughly proportional to the current, the output torque is limited to a value determined by the converter current limit setting.

Consequently, the starting torque is mainly limited by the current limit setting of the converter. It is not economical, and usually not necessary, to design a converter with an excessively high current rating. So the starting torque capability is dependent on the extent to which the converter current rating exceeds the motor rated current. The converter is usually designed to run continuously at its rated current I_N, with an over-current rating of 150% the converter current rating, but for a limited time, usually of 60 sec. The *current limit control* is then set at the 150% level with a protection timer which times out after the period of 60 secs.

Starting torque of the variable speed drive system:

$$T_S = \frac{1.5 \times I_{Convr}}{I_{Motor}} T_N \quad \text{Nm}$$

where T_N = Rated torque of the motor in Nm

Clearly, with an over-sized converter, there is a limit to how much torque the motor will produce above its rated torque. The motor will usually stall at 2.5 to 3 times its rated torque, depending on the design.

For very high starting torques, a larger motor and converter should be considered or the matter should be referred to the manufacturer.

7.7 The requirements for stopping

When a drive is operating in the *first or third quadrants*, the machine is operating as a motor and is driving a mechanical load respectively in the forward or the reverse direction. Energy conversion is taking place from electrical energy to mechanical energy. Energy is stored in the rotating system as **kinetic energy**.

When the drive changes its operation to the *second or fourth quadrants*, braking is required to retard the speed of the mechanical load. To reduce speed, the kinetic energy needs to be removed from the rotating system (AC induction motor plus mechanical load) and transformed into some other form of energy before the system can come to standstill. This is usually a major problem with high inertia loads.

There are several methods of decelerating and stopping a variable speed drive system:

- **Coast to stop**, where the kinetic energy is dissipated in the load itself
- **Mechanical braking**, where the kinetic energy is converted to heat due to friction
- **Electrical braking**, where the kinetic energy is initially converted to electrical energy before being transferred back to the power supply system or dissipated as heat in the motor or a resistance

Most fixed and variable speed drives are stopped by removing the power and allowing the driven machine to coast to a stop. The rotating system, comprising the motor and load, would come to a stop after a 'natural' deceleration t_N time shown in Figure 7.10. This type of stopping is adequate for most mechanical loads such as conveyors, screw conveyors, fans, etc. The actual stopping time depends on the load inertia, load losses and the type of process. However, there are some applications where additional braking is required to provide a shorter deceleration time as shown in Figure 7.10.

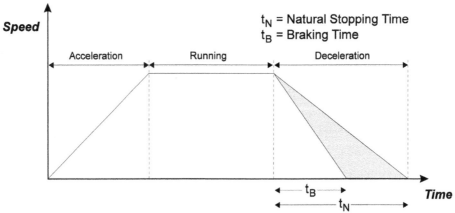

Figure 7.10:
Braking times for rotating drives

The traditional approach was to use mechanical braking, but this requires considerable maintenance to both the mechanical parts and the brake pads. With mechanical braking, the braking energy is dissipated in the form of heat by the friction between the brake pads and the brake disk/drum.

For modern variable speed drive systems, electrical braking is the preferred method of braking. Electrical braking systems rely on temporarily using the motor as an induction generator with the mechanical load driving the generator.

It should be appreciated that, a motor always puts out a torque in a direction so as to cause the rotor to approach synchronous speed of the rotating air-gap magnetic field.

- In the **motoring mode**, the inverter output frequency will always be higher than the rotor speed
- In the **generating (braking) mode**, the inverter output frequency will be at a frequency lower than the rotor speed. The braking torque produced during deceleration is dependent on the slip in the motor.

During electrical braking, energy conversion takes place from mechanical energy to electrical energy. This energy can be disposed of in three ways:

- Dissipated as heat in the rotor of the motor, *DC braking*
- Dissipated as heat in the stator of the motor, *flux braking*
- Dissipated as heat in an external resistor, *dynamic braking*
- Returning electrical energy to the supply, *regenerative braking*

Electrical braking has several advantages over mechanical braking:

- Reduction in the wear of mechanical braking components
- Speed can be more accurately controlled during the braking process
- Energy can sometimes be recovered and returned to the supply
- Drive cycle times can be reduced without any additional mechanical braking

Current-source inverters (CSI) are capable of regenerative braking without modification and other braking techniques need not be considered.

Voltage-source inverters (VSI), which include PWM types, cannot regenerate without costly modifications to the rectifier module. The other electrical braking methods should always be considered first, provided that the cost of the lost energy is not critical.

7.7.1 DC injection braking

The basic principle of *DC injection braking* is to inject a DC current into the stator winding of the motor to set up a stationary magnetic field in the motor air-gap. This can be achieved by connecting two phases of the induction motor to a DC supply. The injected current should be roughly equal to the excitation current or no-load current of the motor.

In a PWM type VVVF converter, DC injection braking is relatively easy to achieve. The inverter control sequence is modified so that the IGBTs in one phase are switched off while the other two phases provide a PWM (pulsed) output to control the magnitude and duration of the DC current. The configuration is shown in Figure 7.11.

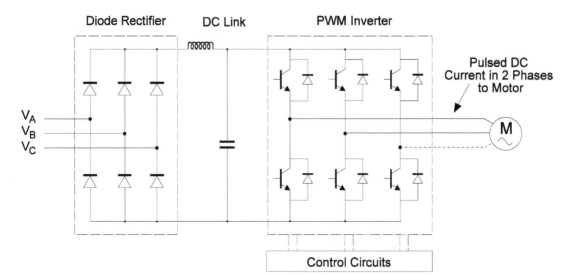

Figure 7.11:
DC injection braking from a PWM converter

As the rotor bars cut through this field, a current will be developed in the rotor with a magnitude and frequency proportional to speed. This results in a braking torque that is proportional to speed. The braking energy is dissipated as losses in the rotor windings, which in turn, generate heat. The braking energy is limited by the temperature rise permitted in the motor. Precautions should be taken to check the motor heating time constant when using this method.

The braking torque will not be high unless the rotor has been designed to give a high starting torque, ie has a high resistance or shows significant deep bar effect.

Another difficulty is that the braking torque will reduce as zero speed is approached and mechanical brakes might be necessary to bring the motor to rest sufficiently quickly or to hold a position at standstill. Nevertheless, the method can still give significant reductions in mechanical brake wear. All the braking power goes into heating up the rotor and this may limit the braking duty.

7.7.2 Motor over-flux braking

A technique that is gaining increasing popularity with modern PWM AC drives is the control of the motor flux. By increasing the inverter output V/f ratio during deceleration, the motor can be driven into an over-fluxed condition, thereby increasing the losses in the motor. The braking energy is then dissipated as heat in the stator winding of the motor. In many ways this is similar to DC injection braking because the braking energy is dissipated in the motor rather than the converter.

Braking torques of up to 50% of rated motor torque are possible with this technique. Again the braking energy is limited by the temperature rise permitted in the motor.

7.7.3 Dynamic braking

When the speed setting of the VVVF converter is reduced, the output frequency supplied to the connected motor is also reduced and the synchronous speed of the motor will decrease. However, this does not necessarily mean that the actual speed of the motor will change immediately. Any changes in the actual motor speed will depend on the external mechanical factors, particularly the inertia of the rotating system.

Figure 7.12 illustrates the change in the motor torque when the converter output frequency is suddenly reduced from f_0 to f_1. The slip changes from being positive (motoring) to negative (generating) and the direction of energy flow is reversed, kinetic energy is converted to electrical energy in the motor, which is then transferred from the motor to the converter.

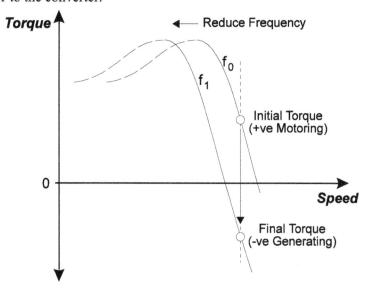

Figure 7.12:
Torque–speed characteristics of an induction motor when frequency is reduced from f_0 to f_1

In practice, the output frequency from the VVVF converter is reduced slowly to avoid the large braking currents that would otherwise flow. For maximum braking torque, the current can be controlled to remain at or below the current limit of the inverter bridge. This procedure can be viewed as being the opposite of the normal startup sequence, where the motor accelerates from standstill to full speed at current limit.

During braking, the converter must have some means of dealing with the energy transferred from the motor. Since the polarity of the DC bus voltage does not change in the braking mode, the direction of the DC bus current reverses during braking. On PWM converters, which use a diode rectifier bridge, the braking current is blocked from returning to the mains power supply. Therefore, unless some mechanism is provided to absorb this energy during braking, the voltage on the DC bus will rise to destructive levels.

With **dynamic braking**, the braking energy is dissipated in a **braking resistor** connected across the DC bus of the converter. As described above, braking is achieved by reducing the inverter output frequency to be less than the actual rotor speed. The slip can be optimized to give as high a torque per ampere as for motoring.

Power flow is from the motor back through the inverter to the DC bus. The braking energy cannot be returned to the mains supply because the input rectifier can only transfer power in one direction. Instead, the energy is absorbed by the DC capacitor, whose voltage rises. To prevent the DC bus voltage rising to a dangerously high level, the capacitor needs to be periodically discharged. This is done by means of a dynamic brake module shown in Figure 7.13, consisting of a power electronic switch, usually an IGBT or BJT, and a discharge resistor connected across the DC link capacitor.

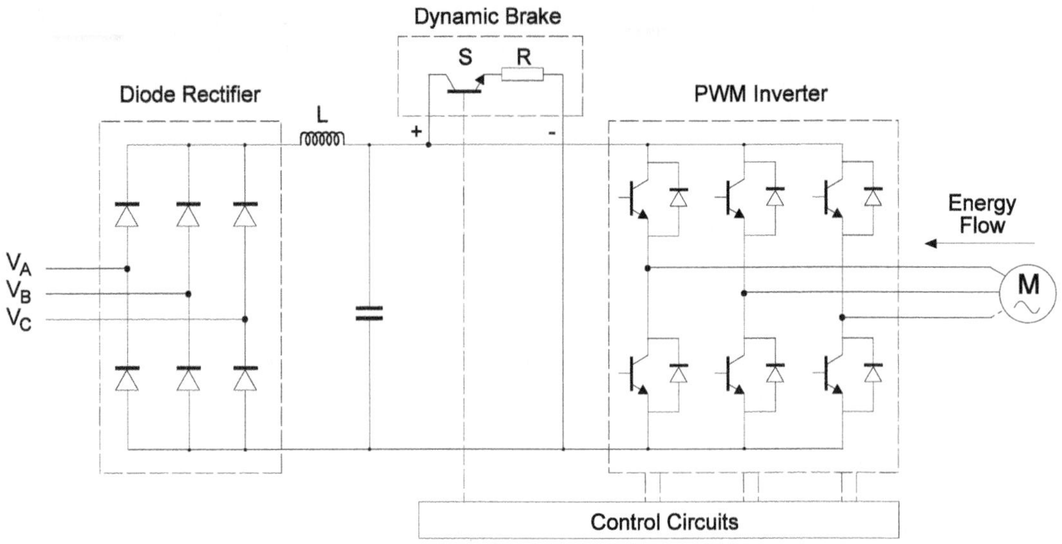

Figure 7.13:
PWM AC converter with a DC link dynamic brake

The IGBT or BJT is controlled by a hysteresis circuit to turn on when the capacitor voltage is too high and turn off when the voltage drops below a certain level, as shown in Figure 7.14. Alternatively, the IGBT may be switched on and off at constant frequency, with duty cycle varying linearly between 0% and 100% as bus voltage changes over specific range.

The switching level of the braking IGBT should be chosen to be higher than the mains supply when it is operating at highest voltage tolerance, but below the maximum safe switching voltage of the inverter components. In practice, for a converter connected to a 3-phase 415 volt supply, with a nominal DC peak voltage of 586 V, the switching level would have to be set at least 10% above this at 650 V, but below 800 V, which is the maximum safe operating voltage of the DC bus. A practical switch-on level is typically 750 V, with the hysteresis between the upper and lower level being 20 V to 30 V lower. The range of allowable voltage swing is determined by the IGBT and capacitor voltage ratings and the tolerance on the supply voltage.

When the motor speed is very low, the inverter can apply a frequency which is slightly negative to maintain the necessary slip for good braking torque, allowing the motor to be electrically braked right down to zero speed. However, this operation requires a high quality of control and braking is often only available to about 2% of rated speed with standard drives.

The resistor value is selected to allow a DC bus current which corresponds to 100% rated torque at maximum motor speed and its power rating must reflect the required braking *duty* with respect to duration, magnitude and frequency of braking. The braking IGBT must be selected to switch the maximum braking current, determined by the value of braking resistor and the maximum bus voltage. The IGBT is usually of the same type and size as that used in the inverter stage.

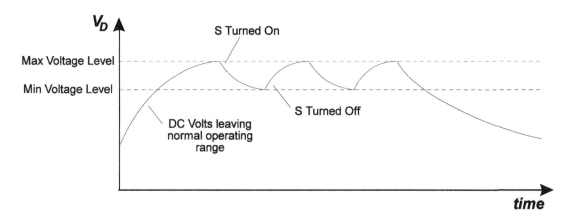

Figure 7.14:
DC bus voltage with hysteresis control on a dynamic brake

Example:

> A 22 kW VSD and motor combination must provide 100% rated motor torque while braking. The maximum braking duty is 3 seconds in every 10 seconds. Assume that the bus voltage during braking is 650 V DC and the DC bus over-voltage trip level is set at 700 V DC. What braking resistor should be used for the application?
>
> To achieve rated torque while braking, the resistor must absorb a full 22 kW when the motor is at full speed. Therefore, the maximum DC bus current will be roughly
>
> $$DC\ Current = \frac{22 \times 10^3}{650} = 34 \quad \text{amps}$$
>
> To absorb 34 amps when the bus voltage is at 650 V DC, the braking resistor will need to have a resistance of 650/34 = 19 ohms. If braking only occurs for 3 sec in every 10 sec, then the duty will be 30%. The power rating of the braking resistor will then be 30% of 22 kW, which is approximately 7 kW continuous. Care must be taken to allow adequate excess power rating when selecting a braking resistor, as the instantaneous power is very high and hot spots can cause premature failure.
>
> The maximum transistor current will occur at the maximum DC bus voltage
>
> $$I_{Max} = \frac{700}{19} = 37 \quad \text{amps}$$
>
> Allowing a safety margin, a braking transistor rated at 50 amps would be selected.

7.7.4 Regenerative braking

From the point of view of the inverter, *regenerative braking* is achieved in a similar way as for dynamic braking. When braking is required, the output frequency of the inverter is reduced to a level below the actual rotor speed. The path for the braking power is from the motor through the reverse connected diodes of the inverter, into the DC bus capacitor, which rises in voltage. Since the normal diode rectifier cannot return power to the mains supply, a thyristor converter must be used.

The two alternative methods are illustrated below:

- If a thyristor rectifier bridge is used in place of the diode bridge to supply power for normal motoring, the current flow in a thyristor rectifier cannot change direction for braking. Regeneration is only possible by changing the polarity of the DC bus voltage. This can be achieved by fitting a reversing switch between the rectifier and the capacitor and switching it according to the required power flow direction. Such a system is useful in drives where braking is occasional rather than continuous and the changeover does not need to be fast (eg small electric locomotives).

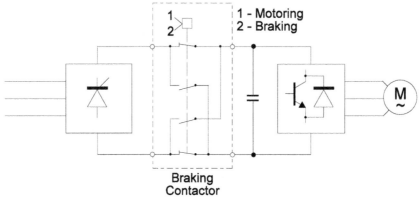

Figure 7.15:
VSD with reversing switch on DC bus for regenerative braking

- For faster transfer to braking, the system in Figure 7.16 can be used with a diode rectifier to supply the motoring power and a thyristor rectifier to extract the braking power.

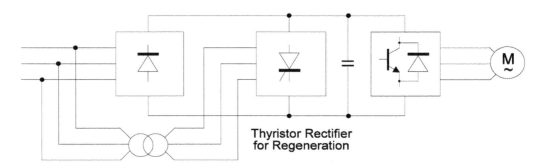

Figure 7.16:
VSD with separate thyristor bridge for regenerative braking

Note that it is not possible to operate both rectifiers from the same level of AC voltage. Suppose the supply voltage is 415 V phase-to-phase. During motoring, from the formula in Chapter 3, the DC bus voltage across the capacitor can be estimated as

$$V_{DC} = 1.35 \times 415 = 560 \quad \text{volts DC}$$

The capacitor voltage will rise during braking with 700 V DC being a typical value. The thyristor rectifier will need to operate as an inverter with a firing angle greater than 90° and with a negative DC voltage, which is the reason for its connection in reverse polarity compared to the diode rectifier.

At firing angles near 180°, slight noise on the supply can prevent a thyristor from fully recovering its forward blocking ability. The firing of the next thyristor gives a short-circuit path across the DC side, a condition called inversion failure and which is difficult to clear. To prevent this, it is usual not to operate a thyristor rectifier with a firing angle greater than 150°.

From the motor point of view, there is essentially no difference between dynamic braking and regenerative braking. The main trade-off is between the initial cost of the regenerative system compared to the economy of returning the braking energy to the mains. This depends on the type of application, the braking effort and the duration of the required braking.

7.8 Control of speed, torque and accuracy

In most VSD applications, simple open-loop speed control without a high degree of accuracy is quite adequate. In these cases, the speed can be set manually or from a PLC and adjusted when required.

In most VSD applications, closed-loop control of the speed is achieved on the basis of a feedback loop from a process variable (PV), such as fluid flow rate in pumping systems. In these systems, the speed does not need to be controlled very accurately because the control system continuously adjusts the speed to meet the process requirements.

But there are some applications where the speed needs to be accurately controlled. In these cases, the following should be considered:

- Does the motor, which has been selected according to the *thermal considerations*, provide sufficient speed accuracy? i.e. Is normal motor slip acceptable?
- It may be advisable to select a larger motor and converter in order to reduce the slip and improve the speed accuracy.
- It may also be necessary to use speed feed-back from the motor, for example a tachometer or digital speed encoder, to obtain accurate speed control. This is called closed-loop speed control?

7.9 Selecting the correct size of motor and converter

Manufacturers of electric motors and frequency converters have evolved various methods for quickly selecting the size of motors and frequency converters for a particular machine load. The same basic procedure is used by most applications engineers. These days, applications selections are usually done on the basis of PC based software. However, it is important for engineers to clearly understand the selection procedure.

One of the best procedures uses a simple *Nomogram* based on the load limit curves to make the basic selection of motor size. This procedure is described below.

The other factors are then checked to ensure that the optimum combination of motor and converter is selected.

The following selection principles are recommended:

1. First, the type and size of motor should be selected. The number of poles (basic speed) should be chosen so that the motor runs as much as possible at a speed slightly above the base speed of 50 Hz. This is desirable because:

- The thermal capacity of the motor improves when f ≥ 50 Hz because of more efficient cooling at higher speeds.
- The converter commutation losses are at minimum when it is operating in the field weakening range above 50 Hz.
- For a constant torque load, a larger speed range is obtained when the motor operates well in the field weakening range at the maximum speed. This means that the most efficient use is made of the torque/speed capability of the variable speed drive. This could mean cost savings in the form of a smaller motor and converter.
- Although many manufacturers claim that their converters can produce output frequencies of up to 400 Hz, these high frequencies are of little practical use except for very special (and unusual) applications. The construction of standard cage motors and the reduction of the peak torque capability in the field weakening zone, restrict their use at frequencies above 100 Hz. The maximum speed at which a standard squirrel cage motor can be run should always be checked with the manufacturer, particularly for larger 2-pole (3000 rev/m) motors of more than 200 kW. The fan noise produced by the motor also increases substantially as the speed of the motor increases.
- A comparison of the torque produced by a 4 pole and a 6 pole motor is shown in Figure 7.17. This illustrates the higher torque capability of the 6 pole machine.

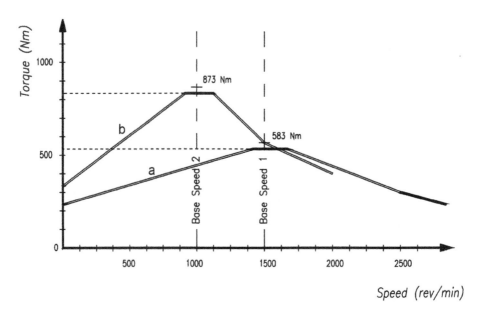

Figure 7.17:
Comparison of the thermal capacity limit curves for two 90 kW TEFC squirrel cage motors

a) 90 kW 4 pole motor (1475 rev/min)

b) 90 kW 6 pole motor (985 rev/min)

2. The selection of an oversized motor just to be 'safe' is not usually advisable because it means that an oversized frequency converter must also be selected. Frequency converters, particularly the PWM-type, are designed for the highest peak current value, which is the sum of the fundamental and harmonic currents in the motor. The larger the motor, the larger the peak currents. To avoid this peak current exceeding the design limit, a converter should never be used with a size of motor larger than for which it is specified. Even when the larger motor is lightly loaded, its harmonic current peaks are high.

3. Once the motor has been selected, it is reasonably easy to select the correct converter size from the manufacturer's catalogue. They are usually rated in terms of current (not kW) based on a specific voltage. This should be used as a guide only, because converters should always be selected on the basis of the maximum continuous motor current. Although most catalogues are based on the standard IEC motor power ratings (kW), motors from different manufacturers have slightly different current ratings.

4. Although it seems obvious, the motor and converter should be specified for the power supply voltage and frequency to which the variable speed drive is to be connected. In most countries using IEC standards, the standard supply voltage is 380 volts ±6%, 50 Hz. In Australia, this is 415 V ±6%, 50 Hz. In some applications where the size of the drive is very large, it is often economical to use higher voltages to reduce the cost of cables. Other commonly used voltages are 500 V and 660 V. In recent years, AC converters are manufactured for use at 3.3 kV and 6.6 kV. Frequency converters are designed to produce the same output voltage as that of the supply, so both the motor and the converter should be specified for the same base voltage. Although the output frequency of the converter is variable, the input frequency (50 Hz or 60 Hz) should be clearly specified because this can have an effect on the design of inductive components.

7.10   Summary of the selection procedures

The selection procedure may now be summarized as follows:

STEP 1.   Specify the initial data for the drive application

To select the correct motor/converter combination, the following information must be available:

- Voltage and frequency of the power supply (volts)
- The breakaway or starting torque (newton meters)
- The load torque (newton meters) and its dependence on speed.
- Speed range of the variable speed drive (rev/min)
- Acceleration requirements or 'ramp times'
- The moments of inertia of the motor and load (kgm^2)

STEP 2.   Selecting the number of poles of the motor

The number of poles determines the synchronous speed of the motor and this is usually selected according to the maximum speed required by the application. Modern VVVF converters are available with output frequencies of up to 400 Hz, although, as pointed out above, there are few practical applications above 100 Hz.

Above-synchronous speeds are of particular advantage for constant torque loads, where the maximum speed should, ideally, be in the range of 50–100 Hz.

This is not the case for pump and fan drives, where the load torque increases as the square of the speed. The optimum use of the motor's torque characteristics occurs when the motor speed is chosen so that the maximum speed of the drive occurs at 50 Hz.

STEP 3. Selecting the motor power rating

Using the load torque requirements, the power rating of the motor can be selected from a motor manufacturer's catalogue using the formula

$$Power = \frac{Torque(\text{Nm}) \times Speed(\text{rev/m})}{9550} \quad \text{kW}$$

However, the de-rating of the motor for harmonic heating, reduced cooling at lower speeds and reduced torque at higher speeds must be taken into account. It is quick and convenient to use a *motor selection nomogram*, an example of which will be discussed at the training course. This nomogram makes allowance for the harmonic heating and reduced cooling of the motor when used with a VVVF converter.

The procedure for using the nomogram is as follows:

- In quadrant 1, first select the column corresponding to the number of poles (synchronous speed) of the selected motor.
- Then select the maximum speed in rev/min of the required speed range. The corresponding frequency can be read off the scale on the right side of quadrant if required.
- Trace horizontally into quadrant 2 up to the load limit curve. The per unit value of torque limit can be read off the scale at the top of the quadrant if required.
- From the intersection of the horizontal trace and the load limit curve, trace vertically downwards into quadrant 3 up to the line corresponding to the calculated load torque. The slope of this curve corresponds to the formula in 3 above.
- From the intersection of the vertical trace and torque line, trace horizontally left to the motor power scale corresponding to the chosen number of poles. The intersection of the horizontal trace and the power scale gives the required motor power in kW.
- For a *square-law* torque load (pump or fan drive), select the standard motor corresponding to the motor power rating.
- For a constant torque load, repeat the above steps to determine the motor power for the minimum speed. Select the standard motor corresponding to the larger of the two power ratings.

STEP 4. Select a suitable frequency converter

A converter with a rating suitable for the motor selected should then be selected from the manufacturer's catalogue. Converters are usually manufactured for power ratings that match the standard sizes of squirrel cage motors. Catalogues usually give the current rating as well as a check to ensure that the motor current is below that of the converter. The following factors must be considered:

- Supply voltage and frequency

- Rated current of the motor
- Duty type (variable torque or constant torque)

A converter is selected so that the rated current of the converter is higher than the rated current of the motor. Also, the type of converter should be suitable for the duty required. Some manufacturers have different converters for the two duty types.

STEP 5. Final checks

The following final checks should be made:

- Is the continuous power rating of the motor (de-rated for altitude, temperature, harmonics, etc) greater than the continuous power requirements of the load?
- Is the starting torque capability of the variable speed drive high enough to exceed the breakaway torque of the load?
- If the VSD is operating in the over-synchronous speed area, is the motor torque capability at maximum speed adequate for the load torque?
- Is the speed accuracy adequate for the application?

STEP 6. An example of a selection calculation

A variable speed drive application has been proposed for a crusher feed conveyer on a mineral processing plant. The required speed range is 600 rev/m to 1400 rev/min. The calculated power requirement of the load, reduced to the motor shaft, is 66 kW at 1400 rev/m. The breakaway torque is expected to be 110% of rated torque. The supply voltage is 415 volts, 50 Hz. Select the optimum size and rating of squirrel cage motor and converter for the most cost effective solution.

The load is a typical constant torque load type. From the previous equations, the constant load torque requirement across the speed range is:

$$T_M = \frac{9550 \times P_A(\text{kW})}{Speed(\text{rev/m})} \quad \text{Nm}$$

$$T_M = \frac{9550 \times 66}{1400} = 450 \quad \text{Nm}$$

Breakaway torque required

$$T_B = 1.1 \times 450 = 495 \quad \text{Nm}$$

There are two alternative solutions for this application

- Consider a 4 pole motor with a rated speed of 1480 rev/min
  Using the motor selection nomogram and plotting the torque requirements for the minimum and maximum speeds, a 110 kW, 415 V, 4-pole motor should be selected based on the motor load capacity at the minimum speed of 600 r/min. (A 75 kW motor would have been chosen on the basis of 1400 rev/m.) This motor frame size is 280 M and current rating 188 amps.

From the motor catalogue, the rated torque of this motor is 710 Nm. The VSD can deliver current equivalent to 150% torque at starting, so the ability to overcome the breakaway torque is generous.

Therefore, the recommended converter is a 110 kW, 415 volt, 220 amp unit.

The rating of the motor and the converter could be decreased if the motor was forced cooled by a separately powered fan.

- Investigate a 6 pole motor with a rated speed of 985 rev/min
  Using the motor selection nomogram and plotting the torque requirements for the minimum and maximum speeds, a 75 kW, 415 volt, 6-pole motor should be selected based on the motor **load capacity** at both the minimum speed of 600 rev/m and maximum speed of 1400 rev/m. The motor frame size is 280 M and current rating 135 amps.

From the motor catalogue, the rated torque of this motor is 727 Nm. The VSD can deliver current equivalent to 150% torque, so the ability to cover the breakaway torque at starting is adequate.

Therefore, the recommended converter is a 75 kW, 415 volt, 140 amp unit.

The second alternative is the most economical solution because, although the cost of both motor alternatives will be roughly the same (same frame sizes = approx same cost), the converter required for the second alternative will have a lower initial cost, because it has a lower current rating.

8

Installation and commissioning

8.1 General installation and environmental requirements

Modern power electronic AC VVVF converters, which are used for the speed control of electric motors, are usually supplied as *stand-alone* units with one of the following configurations. The first two are the most common configurations.

- **IP00 rating**
  Designed for chassis mounting into the user's own enclosure, usually as part of a motor control center (MCC).

- **IP20/IP30 rating**
  Designed for mounting within a '*clean environment*', such as a weatherproof, air-conditioned equipment room. The environment should be free of dust, moisture and contaminants and the temperature should be kept within the specified limits.

- **IP54 rating**
  Designed for mounting outside in a partially sheltered environment, which may be dusty and/or wet.

8.1.1 General safety recommendations

The manufacturer's recommendations for installation should be carefully followed and implemented. The voltages present in power supply cables, motor cables and other power terminations are capable of causing severe electrical shock.

In particular, the local requirements for safety, which is outlined in the wiring rules and other codes of practice should always take priority over manufacturer's recommendations. The recommended safety earthing connections should always be carefully installed before any power is connected to the variable speed drive equipment.

AC variable speed drives have **large capacitors** connected across the DC link as described in Chapter 3. After a VSD is switched off, a period of several minutes must be allowed to elapse before any work commences on the equipment. This is necessary to allow these internal capacitors to fully discharge. Most modern converters include some form of visual indication when the capacitors are charged.

8.1.2 Hazardous areas

In general, power electronic converters should not be mounted in areas which are classified as *hazardous areas*, even when connected to an 'x' rated motor, as this may invalidate the certification. When necessary, converters may be mounted in an approved enclosure and certification should be obtained for the entire VSD system, including both the converter and the motor.

8.1.3 Environmental conditions for installation

The main advantage of an AC variable speed drive (VSD) is that the TEFC squirrel cage motor is inherently well protected from poor environmental conditions and is usually rated at IP54 or better. It can be reliably used in dusty and wet environments.

On the other hand, the AC converter is far more sensitive to its environment and should be located in an environment that is protected from:

- Dust and other abrasive materials
- Corrosive gases and liquids
- Flammable gases and liquids
- High levels of atmospheric moisture

When installing an AC converter, the following environmental limits should be considered:

- Specified ambient temperature: $\leq 40°C$
- Specified altitude: ≤ 1000 metres above sea level
- Relative humidity: $\leq 95\%$

8.1.4 De-rating for high temperature

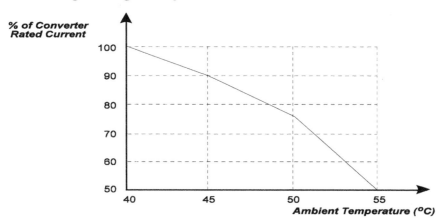

Figure 8.1:
Typical temperature de-rating chart for PWM converter

In regions or environments where there is a high ambient temperature above the accepted 40°C specified in the standards, both the motor and the converter need to be de-rated, which means that they can only be run at loads that are less than their 40°C rating to avoid thermal damage to the insulation materials.

The manufacturers of AC converters usually provide de-rating tables for high temperature environments that are above 40°C. A typical table is given below for a modern PWM converter. This table should be used as a guide only and should NOT be taken to apply to AC converters in general or any converter in particular. The design of AC converters is different from various manufacturers, so the cooling requirements are never the same. The cooling requirements of different models from the same manufacturer may also be different.

8.1.5   De-rating for high altitude

At *high altitudes*, the cooling of electrical equipment is degraded by the reduced ability of the air to remove the heat from the motor or the heat-sink of the converter. The reason is that the air pressure falls with increased altitude, air density falls and, consequently, its thermal capacity is reduced.

In accordance with the standards, AC converters are rated for altitudes up to 1000 meters above sea level. Rated output should be de-rated for altitudes above that.

The manufacturers of AC converters usually provide de-rating tables for altitudes higher than a 1000 m. A typical table is given below for a modern IGBT-type AC converter. Note that this table is NOT applicable to all AC converters. The de-rating of converters with high losses, such as those using BJTs or GTOs, will be much higher than the de-rating required for low loss IGBT or MOSFET converters. The higher efficiency of the latter requires less cooling and would therefore be less affected by altitude changes.

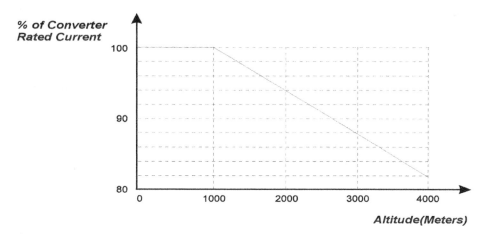

Figure 8.2:
Altitude de-rating chart for IGBT-type converter (Compliments of Allen-Bradley)

8.2   Power supply connections and earthing requirements

In accordance with accepted practice and the local, power is normally provided to a VSD from a distribution board (DB) or a motor control center (MCC). Adequate arrangements should be made to provide safety isolation switches and short-circuit protection in the connection point to the power supply. The short-circuit protection is required to protect the power cable to the AC converter and the input rectifier bridge at the converter. The converter provides down-stream protection for the motor cable and the motor itself.

Adequate safety earthing should also be provided in accordance with the local *wiring rules* and *codes of practice*. The metal frames of the AC Converter and the AC motor should be earthed as shown in Figure 8.3 to keep touch potentials within safe limits. The chassis of the AC converter is equipped with one or more protective earth (PE) terminals, which should be connected back to common safety earth bar.

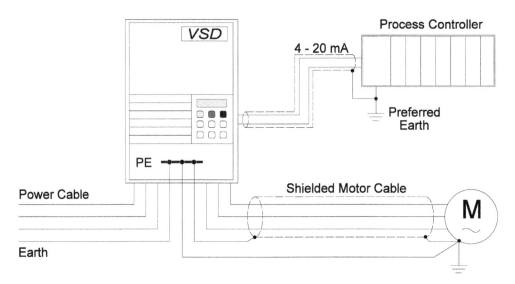

Figure 8.3:
Power supply, motor and earthing connections

8.2.1   Power supply cables

The variable speed drive should be connected to the power supply by means of a cable that is adequate for the current rating of the VSD. Reference can be made to Australian standard AS 3008 when selecting cables. The AC converter requires a 3-phase supply cable (red/white/blue) and a protective earth conductor (green/yellow), which means a 4-core cable with copper or aluminum conductors. A neutral conductor is not necessary and is usually not brought to the frequency converter.

The AC converter is a source of harmonic currents that flow back into the low impedance of the power supply system. This conducted harmonic current is carried into other electrical equipment, where it causes additional heat losses and interference. Sensitive electronic instrumentation, such as magnetic flow-meters, thermocouples and other microprocessor based equipment, ideally should not be connected to the same power source, unless via a filtered power supply.

Also, interference can be radiated from the power supply cable and coupled into other circuits, so these cables should be routed well away from sensitive control circuits.

The power supply cable should preferably be laid in a metal duct or cable ladder and shielded in some way to reduce the radiation of EM fields due to the harmonic currents. Steel wire armored (SWA) cables, are particularly suitable for this purpose. If the power cable is unshielded, control and communications cables should not be located within about 300 mm of the power cable.

The conductor sizes should be selected in accordance with normal economic cable selection criteria, which take into account the maximum continuous current rating of the VSD, the short-circuit rating, the length of the cable and the voltage of the power supply system. The relevant local safety regulations should be strictly observed.

However, when selecting the cable cross-sectional area for the power supply cables and upstream transformers, a de-rating factor of at least 10% should be included to accommodate the additional heating due to the conducted harmonic currents (Chapter 4). If a supply side harmonic filter is fitted at the converter, this may not be necessary. Three-phase systems composed of three single-conductor cables should be avoided if possible. Power cables with a trefoil configuration produce a lower radiated EM field.

8.2.2 Cables between converter and motor

The cable from the AC converter to the motor carries a switched PWM voltage, which is modulated at high frequency by the inverter. This results in a higher level of harmonics than the power supply cable. Harmonic frequencies are in the frequency spectrum of 100 kHz to 1 MHz. The motor cable should preferably be screened or located inside a metal duct. Control and communications cables should not be located close to this cable. The level of radiated EM fields is higher for cables with 3 separate single cores, laid horizontally on a cable ladder, than a trefoil cable with a concentric shield.

The recommended size for the cable between the AC converter and the motor should preferably be the same as the power supply cable. The reasons are:

- It will be easier to add a bypass device in parallel with the frequency converter at a later date, using the same cable, cable lugs and connections.
- The load-carrying capacity of the motor cable is also reduced by harmonic currents and additionally by the capacitive leakage currents.

It should be borne in mind that the AC converter VSD provides short-circuit and overload protection for the cable and motor.

A separate earth conductor between the converter and motor is recommended for both safety and noise attenuation. The earth conductor from the motor must be connected back to the PE terminal of the converter and should not be connected back to the distribution board. This will avoid any circulating high frequency currents in the earth system.

When armored or shielded cables are used between the converter and motor, it may be necessary to fit a barrier termination gland at the motor end when the cable is longer than about 50 m. The reason is that the high frequency leakage currents flow from the cable through the shunt capacitance and into the shield. If these currents return via the motor and other parts of the earthing system, the interference is spread over a larger area. It is preferable for the leakage currents to return to the source via the shortest route, which is via the shield itself. The shield or steel wire armor (SWA) should be earthed at both the converter end and to the frame of the motor.

8.2.3 Control cables

The control cables should be provided in accordance with normal local practice. These should have a cross-sectional area of at least 0.5 mm^2 for reasonable volt drop performance. The control and communications cables connected to the converter should be shielded to provide protection from EMI. The shields should be earthed at one end only, at a point remote from the converter. Earthing the shield to the PE terminal of the drive should be avoided because the converter is a large source of interference. The shield should preferably be earthed at the equipment end.

Cables which have an individual screen for every pair provides the best protection from coupled interference.

The control cables should preferably be installed on separate cable ladders or ducts, as far away from the power cables as possible. If control cables are installed on the same cable ladder as the power cables, the separation should be as far as possible, with the minimum distance being about 300 mm. Long parallel runs on the same cable ladder should be avoided.

8.2.4 Earthing requirements

As mentioned earlier, both the AC converter and the motor must be provided with a safety earth according to the requirements of local standards. The main purpose of this earthing is to avoid dangerous voltages on exposed metal parts under fault conditions.

When designing and installing these earth connections, the requirements for the reduction of EMI should also be achieved with these same earth connections. The main earthing connections of an AC converter are usually arranged as shown in Figure 8.3.

The PE terminal on the converter should be connected back to the system earth bar, usually located in the distribution board. This connection should provide a low impedance path back to earth.

8.2.5 Common cabling errors

The following are some of the common cabling errors made when installing VSDs:

- The earth conductor from the AC converter is run in the same duct or cable ladder as other cables, such as control cables and power cables for other equipment. Harmonic currents can be coupled into sensitive circuits. Ideally, instrument cables should be run in separate metal ducts or steel conduit.
- Running unshielded motor cable next to the supply cable to the AC converter or the power cables for other equipment. High frequency harmonic currents can be coupled into the power cable, which can then be conducted to other sensitive electronic equipment. Other cables should be separated from the motor cable or converter power cable by a minimum of 300 mm.
- Running excessively long cables between the AC converter and the motor, these should be no longer than 100 m. If longer cables are necessary, motor filters are necessary to reduce the leakage current. Alternatively, the switching frequency may be reduced.

8.3 Start/stop control of AC drives

The protection requirements for AC variable speed drives is covered in considerable detail in Chapter 5: Protection of AC converters and motors. The protection of the mains supply side of the converter requires short circuit protection either in the form of a set of adequately rated fuses, usually as part of a switch-fuse unit, or a main circuit breaker.

The stop/start control of the AC drive can be achieved in a number of ways, mainly:

- Controlling the start/stop input of the converter control circuit
- Breaking the power circuit by means of a contactor

The first method is the recommended method of controlling the stopping and starting of an AC converter. This may be achieved by stop and start pushbuttons wired directly to the control terminals of the converter as shown in Figure 8.7.

Alternatively, if the control is from a remote device such as a PLC, the control can be wired from the PLC directly to the terminals of the AC converter as shown in Figure 8.8.

The second method is the one most commonly used for the direct on line (DOL) starting of normal fixed speed AC motors. Following from previous DOL *'standard'* practice, this method is also quite commonly used in industry for the control of variable speed drives, particularly for conveyors. It is usually a safety requirement to interrupt the power circuit when an emergency stop or pull-wire switch has been operated. While this method satisfies the safety requirements by breaking the power supply to the motor, there are a number of potential hazards with this method of control. The main problems are:

- **Contactor on supply side of the AC converter**
  Opening/closing the supply side of the AC converter for stop/start control should be avoided because most modern converters take their power from the DC bus. Every time the power is removed

    – Power to the control circuits is lost

    – Control display goes off

    – Diagnostic information disappears

    – DC capacitors become discharged

    – Serial communications are lost

  When the AC variable speed drive needs to be restarted, there is a time delay (typically 2 secs) while the DC bus charging system completes its sequence to recharge the DC capacitor. This stresses the charging resistors, the DC capacitor and other components. The charging resistors of many AC converters are short-time rated and, although sometimes not highlighted in the user manual, there is a limit to the number of starts that can be done. Many users have the concept of *'run on power up'* is acceptable and unrestricted. The following is an extract from the manual of one of the leading manufacturers of AC converters:

  ATTENTION: The drive is intended to be controlled by control input signals that will start and stop the motor. A device that routinely disconnects and then reapplies line power to the drive for the purpose of starting and stopping the motor is not recommended. If this type of circuit is used, a maximum of 3 stop/start cycles in any 5 minute period (with a minimum period of 1 minute rest between each cycle) is required. These 5 minute periods must be separated by 10 minute rest cycles to allow the drive precharge resistors to cool. Refer to codes and standards applicable to your particular system for specific requirements and additional information.

- **Contactor on motor side of the AC converter**
  Opening/closing the 3-phase power circuit on the motor side of the AC converter for stop/start control should also be avoided, particularly while the AC drive is running. Breaking the inductive circuit to the motor produces

transient over-voltages which can damage the IGBTs and other components. Many modern AC converters have RC suppression circuits (snubbers) to protect the IGBTs from this type of switching. The following is an extract from the manual of one of the leading manufacturers of AC converters:

ATTENTION: Any disconnecting means wired to the drive output terminals U, V and W must be capable of disabling the drive if operated during drive operation. If opened during drive operation, the drive will continue to produce output voltage between U, V and W. An auxiliary contact must be used to simultaneously disable the drive or output component damage may occur.

The objective is to ensure that the AC converter is OFF before the contacts between the converter and the motor are opened. This will avoid IGBT damage due to transient over-voltages.

In addition, closing the motor side contactor when converter output voltage is present can result in a motor inrush current similar to DOL starting. Apart from the stress this places on the converter, the drive will trip on over-current. Repeated attempts at closing the motor contactor after the converter has started may eventually lead to IGBT failure.

If a contactor has to be installed into the power circuit of an AC variable speed drive system to meet local safety requirements, then it is better to locate this contactor *downstream* of the AC converter. It is then necessary to include an auxiliary contact on the contactor which disables the converter control circuit before the contactor is opened or, alternatively, closes the enable circuit after the contactor has been closed. This means that a *late make - early break* auxiliary contact should be used on the contactor and wired to the converter enable input.

While the above configuration will protect the AC converter from failure, this method of routine stop/start control is not recommended. It should be used for emergency stop conditions only. Routine stop/start sequences should be done from the AC converter **control terminals**. An alternative method of ensuring that plant operators follow this requirement is to install a latching relay and a *reset* pushbutton. The latching relay needs to be reset after every Emergency Stop sequence.

8.4 Installing AC converters into metal enclosures

If the environmental conditions are likely to exceed these accepted working ranges, then arrangements should be made to provide additional cooling and/or environmental protection for the AC converter. The temperature limits of an AC converter are far more critical than those for an electric motor. Temperature de-rating needs to be strictly applied. However, it is unlikely that a modern PWM converter will be destroyed if the temperature limits are exceeded. Modern AC converters have built-in thermal protection, usually a silicon junction devices, mounted on the heat-sink. The main problem of over-temperature tripping is associated with *nuisance tripping* and the associated downtime.

Although the efficiency of a modern AC converters is high, typically ± 97%, they all generate a small amount of heat, mainly due to the commutation losses in the power electronic circuits. The level of losses depends on the design of the converter, the PWM switching frequency and the overall power rating. Manufacturers provide figures for the losses (watts) when the converter is running at full load. Adequate provision should be made to dissipate this heat into the external environment and to avoid the temperature inside the converter enclosure rising to unacceptably high levels.

Converters are usually air-cooled, either by convection (small power ratings) or assisted by cooling fans on larger power ratings. Any obstruction to the cooling air flow volume to the intake and from the exhaust vents will reduce efficiency of the cooling. The cooling air volume flows and the power loss dissipation determine the air-conditioning requirements for the equipment room.

The cooling is also dependent on there being a temperature differential between the heat-sink and the cooling air. The higher the ambient temperature, the less effective is the cooling. Both the AC converter and motor are rated for operation in an environment where temperature does not exceed 40°C.

When AC converters are mounted inside enclosures, care should be taken to ensure that the air temperature inside the enclosure remains within the specified temperature limits. If not, the converters should be de-rated in accordance with the manufacturer's de-rating tables.

In an environment where condensation is likely to occur during the periods when the drive is not in use, anti-condensation heaters can be installed inside the enclosure. The control circuit should be designed to switch the heater on when the drive is de-energised. The heater maintains a warm dry environment inside the enclosure and avoids moisture being drawn into the enclosure when the converter is switched off and cools down.

AC converters are usually designed for mounting in a vertical position, to assist convectional cooling. On larger VSDs, cooling is assisted by one or more fans mounted at the bottom or top of the heat-sink.

Many modern converters allow two alternative mounting arrangements:

- **Surface mounting**, where the back plane of the converter is mounted onto a vertical surface, such as the back of an enclosure. (Figure 8.4 and 8.5)
- **Recessed mounting**, where the heat-sinks on the back of the converter project through the back of the enclosure into a cooling duct. This allows the heat to be more effectively dissipated from the heat-sinks. (Figure 8.6)

Sufficient separation from other equipment is necessary to permit the unrestricted flow of cooling air through the heat-sinks and across the electronic control cards. A general *rule of thumb* is that a free space of 100 mm should be allowed around all sides of the VSD. When more than one VSD are located in the same enclosure, they should preferably be mounted side by side rather than one above the other. Care should also be taken to avoid locating temperature sensitive equipment, such as thermal overloads, immediately above the cooling air path of the VSD.

Adequate provision must be made to dissipate the converter losses into the external environment. The temperature rise inside the enclosure must be kept below the maximum rated temperature of the converter.

8.4.1 Calculating the dimensions of the enclosure

The enclosure should be large enough to dissipate the heat generated by the converter and any other electrical equipment mounted inside the enclosure. The heat generated inside an enclosure is transferred to the external environment mainly by radiation from the surface of the enclosure. Consequently, the surface area must be large enough to dissipate the internally generated heat without allowing the internal temperature to exceed rated limits.

The surface area of a suitable enclosure is calculated as follows:

$$A = \frac{P}{k(T_{\text{Max}} - T_{\text{Amb}})}$$

where: A Effective heat conducting area in m^2
            (Sum of surface areas not in contact with any other surface)
         P Power loss of heat producing equipment in watts
         T_{Max} Maximum permissible operating temperature of converter in °C
         T_{Amb} Maximum temperature of the external ambient air in °C
         k Heat transmission coefficient of enclosure material

Example:
Calculate the minimum size of an IP54 cubicle for a typical PWM type frequency converter rated at 22 kW. The following assumptions are made:

- The converter losses are 600 watts at full rated load.
- The converter is to be mounted within an IP54 cubicle made of 2 mm steel.
- The enclosure is effectively sealed from the outside and heat can only be dissipated from the enclosure by conduction through the steel and by radiation from the external surface into the outside air.
- The cubicle stands on the floor with its back against the wall in an air-conditioned room with a maximum ambient temperature 25°C.
- The converter can operate in a maximum temperature of 50°C.
- The heat transmission coefficient is 5.5 (typical for painted 2 mm steel).

The first step is to calculate the minimum required surface area of the enclosure. This can be done by applying the formula for surface area.

$$A = \frac{600}{5.5(50 - 25)} = 4.36 \, \text{m}^2$$

If the cubicle is standing on the floor against a wall, this area applies only to the top, front and two sides of the enclosure. A suitable cubicle can be chosen from a range of standard cubicles or could be fabricated for this installation. In either case, it is important to take into account the dimensions of the converter and to ensure that there is at least 100 mm space on all sides of the converter.

With these requirements in mind, the procedure is to choose or estimate at least two of the dimensions and the third can be derived from the above equation. This calculated dimension must then be checked to ensure that the required 100 mm clearance is maintained.

For a cubicle with dimensions H × W × D standing on the floor against the wall, the effective heat conducting area is

$$A = \text{HW} + 2\text{HD} + \text{WD}$$

Assuming that a standard cubicle is chosen with a height of 2.0 m and a depth of 0.5 m, the width is derived from

$$A = 2.0 \text{ W} + 2 + 0.5 \text{ W}$$
$$A = 2.5 \text{ W} + 2$$

Using the required heat dissipation area from the above calculation

$$4.36 = 2.5 \text{ W} + 2$$
or
$$2.5\text{W} = 2.36$$

$$\text{W} = 0.94$$

Based on the requirements of heat dissipation, the width of the cubicle would have to be larger than 0.94 m. In this case a standard width of 1.0 m would be selected.

Clearances around the sides of the converter should be checked. With typical converter dimensions of H × W × D = 700 × 350 × 300, the cubicle chosen would provide more than 100 mm of clearance around all the converter and also leave sufficient space for cabling and other components.

From this calculation, it is clear that the overall dimensions of the cubicle can be reduced by the following changes:

- Standing the cubicle away from the wall, at least 200 mm
- Reducing the ambient temperature, turning down the air-conditioning
- Providing ventilation to the cubicle to improve heat transfer

8.4.2   Ventilation of enclosures

The enclosure can be smaller if some additional ventilation is provided to exchange air between the inside and outside of the cubicle. There are several ventilation techniques commonly used with converters, but they mainly fall into two categories:

- **Natural ventilation**
  Relies on the convectional cooling airflow through vents near the bottom of the cubicle and near the top, the 'chimney' effect.

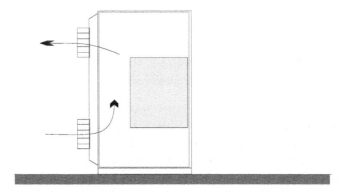

Figure 8.4:
Natural ventilation of a converter in a cubicle

- **Forced ventilation**

  Relies on cooling airflow assisted by a fan located either near the top or the bottom of the cubicle. It is difficult to maintain a high IP rating with ventilated cubicles, so ventilated cubicles need to be located in a protected environment, such as a dust-free equipment room.

  For cooling purposes, a certain volume of airflow is required to transfer the heat generated inside the enclosure to the external environment. The required airflow can be calculated from the following formula:

$$V = \frac{3.1\,P}{(T_{\text{Max}} - T_{\text{Amb}})}$$

where: V = Required airflow in m³ per hour

P = Power loss of heat producing equipment in watts

T_{Max} = Maximum permissible operating temperature of converter in °C

T_{Amb} = Maximum external ambient temperature in °C

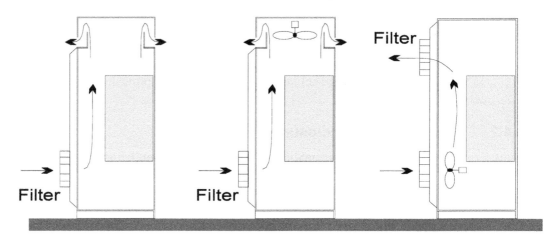

Figure 8.5:
Forced ventilation of a converter in a cubicle

Example:

Calculate the airflow ventilation requirements of the 22 kW converter used in the example above, using the same assumptions.

The required airflow to maintain adequate cooling

$$V = \frac{3.1 \times 600}{(50 - 25)} = 74.4 \quad \text{m}^3/\text{h}$$

An airflow of 75 m³/h is necessary to remove the heat generated inside the enclosure by the converter and to transfer it to the outside. In this case, the dimensions of the cubicle are based purely on the minimum physical dimensions required for the converter and any other equipment mounted in the cubicle.

This airflow could be achieved by the convectional flow of air provided that the size of the top/bottom openings are large enough and the resistance to airflow is not unnecessarily restricted by dust-filter pads. Alternatively, a fan assisted ventilation system would be necessary to deliver the required airflow.

8.4.3 Alternative mounting arrangements

One of the main problems associated with the ventilation of converter cubicles is that it is very difficult to achieve a high IP rating with a ventilated cubicle. In addition, if filters are used, an additional maintenance problem is introduced, the filters need to be checked and replaced on a regular basis.

A solution, which is rapidly gaining popularity, is the **recessed mounting**. This technique has now been adopted by many of the converter manufacturers.

Most of the heat generated by a converter is associated with the power electronic components, such as the rectifier module, inverter module, capacitors, reactor and power supply. These items are usually mounted onto the heat-sink base of the converter and most of the heat will be dissipated from the surfaces of this heat-sink. The digital control circuits do not generate very much heat, perhaps a few watts.

If the heat-sink is recessed through the back mounting plane of the enclosure, most of the heat will be dissipated to the environment external to the cubicle. The portion of the converter with the control circuits remain within the enclosure. With a suitable seal around the converter, the enclosure can be relatively small and rated at >IP54 without the need for forced or convectional airflow ventilation.

The heat-sink portion projecting outside the enclosure can be exposed to the environment with a lower IP rating (e.g. IP20) or it can be arranged to project into a cooling airduct system, which ducts the heat outside the building. Figure 8.6 shows a typical mounting arrangement of this type of converter with the heat-sinks projecting into a cooling duct.

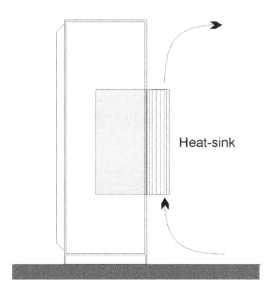

Figure 8.6:
Converter mounted with heat-sink outside the cubicle

8.5 Control wiring for variable speed drives

Variable speed drives (VSD) may be controlled '*locally*' by means of manual push-buttons, switches and potentiometers, mounted on the front of the converter. For simple, manually controlled operations, these local controls are all that is required to operate the VSD.

In most industrial applications, it is not practical to control the VSD from the position where the VSD is located. VSDs are usually installed inside motor control centres (MCCs), which are located in switchrooms, usually close to the power supply transformer, but not necessarily close to where the operator is controlling the process.

Consequently, almost all VSDs have terminals that permit *remote control* from a location close to the operator. VSDs have terminals for the following controls:

- **Digital inputs**, such as remote start, stop, reverse, jog, etc, which are usually implemented by
  - Remote push-buttons in a manually controlled system
  - Digital outputs (DO) of a process controller in an automated system

- **Digital status outputs**, such as indication of running, stopped, at speed, faulted, etc, which are usually implemented by
  - Remote alarm and indication lamps in a manually controlled system
  - Digital inputs (DI) to a process controller in an automated system

- **Analog inputs**, such as remote speed reference, torque reference, etc, which are usually implemented by
  - remote potentiometer (10 kohm pot) in a manually controlled system
  - Analog outputs (AO) of a process controller in an automated system, usually using a 4–20 mA signal carried on a screened twisted pair cable

- **Analog outputs**, such as remote speed indication, current indication, etc, which are usually implemented by
  - Remote display meters (0–10 V) in a manually controlled system
  - Analog inputs (AI) to a process controller in an automated system, usually using a 4–20 mA signal carried on a screened twisted pair cable

Manual and automated control systems have operated very effectively for many years with this type of '*hard-wired*' control system. The main disadvantage of this system is:

- All the DIs and DOs require one wire per function, plus a common.
- All the AIs and AOs require two wires per function, plus a shield connection.

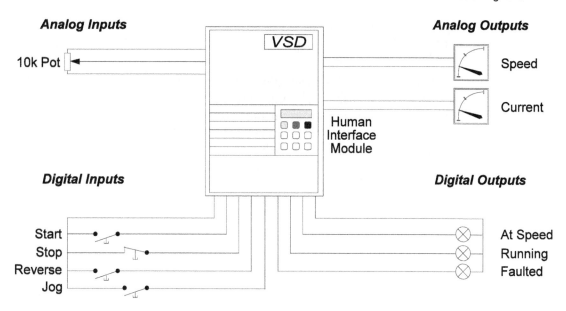

Figure 8.7:
Configuration of a typical hard-wired manual control system

8.5.1 Hard-wired connections to PLC control systems

With the introduction of automated control systems using programmable logic controllers (PLCs) and distributed control systems (DCS), the '*hard-wired*' control connections have been extended, with the input/output (I/O) modules replacing the manual controls.

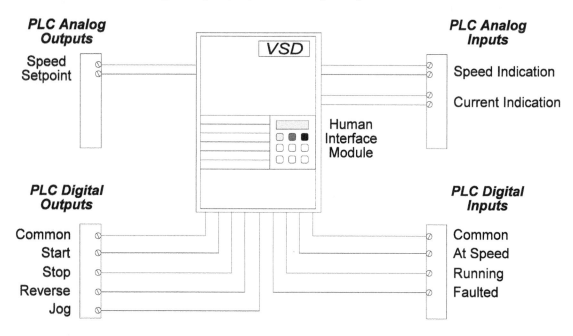

Figure 8.8:
Configuration of a typical hard-wired automated control system

As the control systems have grown in complexity and the amount of information required from field sensors has expanded, the number of conductors required to implement the automated control system has become a major problem, from the point of view of cost and complexity. As more and more field devices become integrated into the overall control system, this problem of more and more cables can only become more difficult.

A hard-wired interface between a variable speed drive (VSD) and a programmable logic controller (PLC) would typically require about 15 conductors as follows:

- 5 Conductors for controls such as start, stop, enable, reverse, etc
- 4 Conductors for status/alarms, such as running, faulted, at speed, etc
- 2 or 3 Conductors for analog control, such as speed setpoint
- 4 Conductors for analog status, such as speed indication, current indication

If there are several VSDs in the overall system, the number of wires is multiplied by the number of VSDs in the system.

8.5.2 Serial communications with PLC control systems

Serial communications overcomes these problems and allows complex field instruments and VSD systems to be more simply linked together into an overall *automated control system* with the minimum of cabling. Microprocessor based digital control devices, sometimes called '*smart*' devices, are increasingly being used in modern factory automation and industrial process control systems. Several '*smart*' devices can be '*multi-dropped*' or '*daisy-chained*' on one pair of wires and integrated into the overall automated control system. Control and status information can be transferred serially between the process controller and the VSDs located in the field. Parameter settings can also be adjusted remotely from a central point.

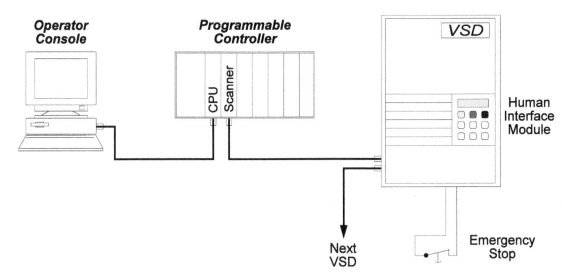

Figure 8.9:
Configuration of a typical serial communications system

Transmitter		EIA-232	EIA-423	EIA-422	EIS-485
Mode of operation		Unbalanced	Unbalanced	Differential	Differential
Max number of drivers & receivers on line		1 Driver 1 Receiver	1 Driver 10 Receivers	1 Driver 10 Receivers	32 Drivers 32 Receivers
Maximum cable length		15 m	1200 m	1200 m	1200 m
Maximum data rate		20 kbps	100 kbps	10 Mbps	10 Mbps
Maximum common mode voltage		±25 V	±6 V	+6 V to −0.25 V	+12 V to −7 V
Driver output signal		±5.0 V min ±25 V max	±3.6 V min ±6.0 V max	±2.0 V min ±6.0 V max	±1.5 V min ±6.0 V max
Driver load		≥3 kohm	≥450 ohm	≥100 ohm	≥60 ohm
Driver output resistance	Power On	n/a	n/a	n/a	≤100 microA −7 V ≤ Vcm≤ 12 V
(High-Z state)	Power Off	300 ohm	≤100 microA @ ±6V	≤100 microA −0.25 V ≤ Vcm≤6 V	≤100 microA −7 V ≤ Vcm≤ 12 V
Receiver input Resistance		3 kohm to 7 kohm	≥4 kohm	≥4 kohm	≥12 kohm

Figure 8.10:
Comparison of features of EIA-232, EIA-423, EIA-422 and EIA-485

This level of the control system is usually called the '*field level*' and a data communications network at this level is referred to as a '*field bus*'.

The *physical interface standards* define the electrical and mechanical details of the interconnection between two pieces of electronic equipment, which transfer serial binary data signals between them. There are a several well established physical interface standards, such as EIA-232, EIA-422 and EIA-485. The main features of the four most common EIA interface standards are compared in the Table in Figure 8.10.

8.5.3   Interface converters

While many PCs and PLCs are fitted with RS-232 interfaces as standard, these ports are not generally suitable for control of variable speed drives directly via this interface, because of differences in both the voltage levels and the configuration (unbalanced/balanced).

An RS-232/RS-422 or RS-232/RS-485 interface converter can be connected between the two devices to convert the voltage levels and the connection configuration from one to the other (unbalanced to balanced). An interface converter should be physically located close to the RS-232 port (at the PC end) to take advantage of the better performance characteristics of the RS-485 interface for the longer distance to the variable speed drives in the field. It is also preferable that these converters and the interface at the drive end *optically isolate* the control device from the variable speed drives to provide maximum protection against problems resulting from earth voltage differences and electrical noise.

The interface between the PC and the interface converter is one-to-one, while the RS-485 side may have several devices (up to 32) connected in a multi-drop configuration. The internal connection details of an RS-232/RS-485 converter are shown in the Figure below.

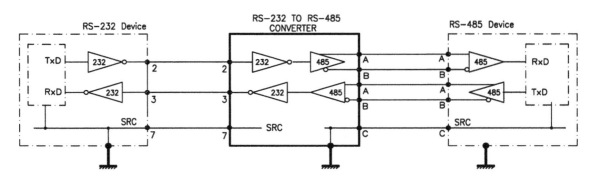

Figure 8.11:
Block diagram of an RS-232/RS-485 converter

8.5.4   Local area networks

The technique of linking digital instrumentation and control devices together to share data and pass on control commands has become an important part of automated industrial control systems. It is fairly common for the main components of the control system to be linked together by a '*data highway*'. Another name for a data highway is a *local area network (LAN)*.

A *LAN* is a communications path linking two or more intelligent devices. A LAN allows shared access by several users to the common communications cable network, with full connectivity between all nodes on the network. A LAN covers a relatively small area and is located within a localized plant or group of buildings.

The connection of a communicating device into a LAN is made through a node. A node is any point where a device is connected and each node is allocated a unique address number. Every message sent on the LAN must be prefixed with the unique address of the destination. A device connected through a node into a network, receives all messages transmitted on the network, but only responds to messages sent to its own address. LANs operate at relatively high speed (50 kbps and upwards) with a shared transmission medium over a fairly small local area.

Since many nodes can access the LAN network at the same time, the network software must deal with the problems of sharing the common resources of the network without conflict or corruption of data. In the OSI model, this level of software is called the *data link layer*. Some rules must be established on which devices can access the network, when and under what conditions.

Most modern digital *variable speed drives (VSDs)* have a communications capability. The physical interface is usually based on a well known physical standard, such as RS-232 or RS-485. There are a number of well accepted industrial communications standards, for example, *Devicenet, Profibus, Asi-bus, Modbus Interbus-s*, etc. A suitable program embedded in the VSD can control the serial transfer of data to the VSD. There is still a considerable amount of confusion about the merits of these emerging new standards, but these will be resolved in the next few years.

8.6 Commissioning variable speed drives

8.6.1 The purpose of commissioning

The main purpose of commissioning VSDs is to ensure that:

- The AC converter and motor have been correctly installed and meet the wiring and safety standards such as AS 3000.
- The power and motor cables are correctly sized, installed and terminated.
- All power cable shields have been correctly earthed at both ends, to the PE terminal at the converter, at the motor and at the DB or MCC.
- The control cables have been installed according to the control system design.
- All control cable shields have been correctly earthed at one end only, preferably at the process control system end ('cleaner' earth).
- There are no faults on the cables prior to energization for the first time.

8.6.2 Selecting the correct application settings

Once all the basic checks have been completed and the commissioning test sheet completed, the VSD is ready for energization.

It is recommended that, when energizing the converter for the first time, that the motor cables should be disconnected until all the basic parameter settings have been installed into the converter. This will avoid problems with starting the motor in the wrong direction, starting with an acceleration time which is too fast, etc. There is no danger in running a PWM converter with the output side completely open circuit.

Once all the initial settings and on-load checks have been completed, the motor cable can be insulation tested and connected for the final on-load commissioning tests.

8.6.3 Selecting the correct parameter settings

A variable speed drive will only perform correctly if the basic parameters have been correctly set to suit the particular application. The following are the basic parameters that must be checked before the VSD is connected to a mechanical load:

- The correct **base voltage** must be selected for the supply voltage and to suit the electric motor connected to the output. In Australia, this standard voltage is usually 415 V, 3-phase. This will ensure that the correct output volts/Hz ratio is presented to the motor.
- The correct **base frequency** must be selected for the supply voltage and to suit the electric motor connected to the output. In Australia, this standard frequency is usually 50 Hz. This will ensure that the correct output volts/Hz ratio is presented to the motor.
- The connections to the cooling fan should be checked to ensure that the correct tap on the transformer has been selected.

Thereafter, the remaining parameters settings can be selected as follows:

- **Maximum speed**, usually set to 50 Hz, but often set to a higher speed to suit the application. Reference should be made to Chapter 6 to ensure that the maximum speed does not take the drive beyond the **loadability limit**.

- **Minimum speed**, usually 0 Hz for a pump or fan drive, but often set at a higher speed to suit constant torque applications. Reference should be made to Chapter 6 to ensure that the minimum speed does not take the drive below the **loadability limit**.

- **Rated current of the motor**, this depends on the size of the motor relative to the rating of the converter. The current rating of the converter should always be equal to or higher than the motor rating. For adequate protection of the motor, the correct current rating should be chosen.

- **Current limit**, determines the starting torque of the motor. If a high breakaway torque is expected, a setting of up to 150% will provide the highest starting torque.

- **Acceleration time**, determines the ramp-up time from zero to maximum speed. This should be chosen in relation to the inertia of the mechanical load and the type of application. For example, in a pumping application, the acceleration time should be slow enough to prevent water hammer in the pipes.

- **Deceleration time**, determines the ramp-down time from maximum speed to zero. This setting is only applicable if the 'ramp to stop' option is selected. Other alternatives are usually 'coast to stop' and 'DC braking'. On high inertia loads, this should not be set too short. If the deceleration time is below the natural rundown time of the load, the DC voltage will rise to a high level and could result in unexpected tripping on '*over-voltage*'. The deceleration time can only be shorter than the natural rundown time if a dynamic braking resistor has been fitted.

- **Starting torque boost**, can be selected if the load exhibits a high *breakaway torque*. This feature should be used cautiously to prevent over-fluxing of the motor at low speeds. Too high a setting can result in motor over-heating. Only sufficient torque boost should be selected to ensure that the VSD exceeds the breakaway torque of the load during starting.

There are also many other settings commonly required on modern digital VSDs. The above are the most important and must be checked before starting. The remaining parameters usually have a 'default' setting which will probably be adequate for most applications. However, these should be checked and adjusted for optimum operation.

9

Special topics and new developments

9.1   Soft-switching

The present design of *PWM voltage source inverters* involves a constant DC link voltage supply and semiconductor switching devices, with anti-parallel diodes, feeding an inductive load. When device switching occurs, the other anti-parallel diode in the same leg conducts and assures that full voltage is across the switching device. This gives it the so-called *clamped inductive load switching waveform*, as shown in Figure 9.1. This leads to the simultaneous large voltages and currents that give rise to high switching losses in an inverter.

A new inverter topology, which is under investigation, gives either **zero voltage** or **zero current** during switching, to reduce switching power loss to a very low level. This new technique is called *soft switching* and should allow future semiconductor devices to be switched at much higher frequencies, thereby giving better waveform control.

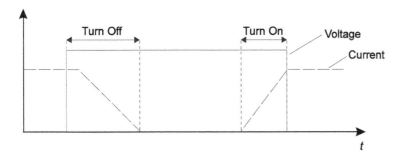

Figure 9.1:
Characteristic of clamped inductive load switching waveform

One of the possible designs for achieving this goal is the resonant link inverter, which is shown for a single-phase case in Figure 9.2. The front-end is identical to a normal '*hard switched*' inverter, except for the series inductor and shunt capacitor between the DC bus and the inverter stage.

The circuit is controlled as follows:

- Assume that the capacitor is momentarily discharged from the previous cycle of inverter operation.
- All inverter switches are turned ON at zero voltage, applying zero volts to the load and shorting out the capacitor. The inductor current then ramps up through the switches.
- When the inductor current reaches an appropriate level, one switch in each leg is opened (at zero voltage) to apply voltage to the load. The capacitor voltage then rings up to a value exceeding the supply while the inductor current decreases. The oscillation continues with capacitor voltage now decreasing.
- When the capacitor voltage decreases to zero, the anti-parallel diodes clamp the capacitor voltage from going negative, in effect placing a short-circuit across the capacitor and momentarily discharging it.
- The process is repeated.

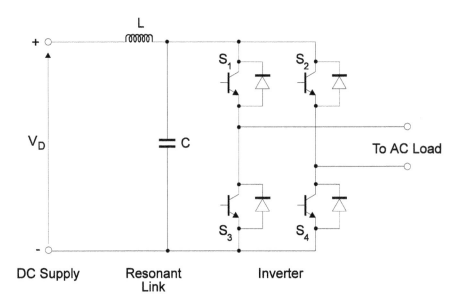

Figure 9.2:
The topology of a single-phase resonant link 'soft' inverter

The supply across the inverter legs has the form of a series of pulses with the same waveform as the capacitor voltage as shown in Figure 9.3. The inverter legs must switch at one of the voltage zeros if the resonance is to be continued and low switching losses are to be achieved.

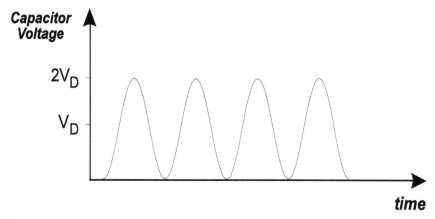

Figure 9.3:
Resonant link inverter waveform at inverter leg

This circuit does not have the ability to give pulses with continuous variation in pulse width as with the conventional inverter. The output voltage must be controlled by discrete pulse modulation rather than pulse width modulation. However, the resonance is at such a high frequency (50–100 kHz) that this does not limit the smoothness of the output current because the load inductance is very effective at filtering such a high frequency.

There are several other types of soft-switching circuits under investigation at present. Some of them do not maintain continuous resonance but are controlled to resonate at a desired moment of switching, which allows continuous pulse width modulation. This type of design, called a *quasi-resonant link inverter*, is related to the force-commutation circuits, except that operation is at much higher frequencies and is used with gate controlled devices which can inherently be turned off. Other designs allow zero-current rather than zero-voltage turn off.

One of the advantages of this type of switching is that it results in lower levels of RFI because of the slower rates of rise of voltage and current.

Most of the potential designs share common problems.

- Resonance inherently causes higher voltages than that of the supply, which places higher stresses on the power switches and the load. This can be overcome with the addition of other switches and energy storage elements to absorb excess energy.
- They require more complex control systems because the instant of switching has to be varied with the load to maintain resonance. This control must be implemented at about 20 times the switching frequency of a conventional inverter. This requires fast controller hardware and software, such as a *digital signal processor*.

9.2 The matrix converter

All AC converter designs discussed in this book so far contain energy storage elements, such as inductance and capacitance, as well as the semiconductor power switches. The energy storage components result in extra losses, are bulky, and contribute to unreliability. The matrix converter attempts to eliminate these storage devices.

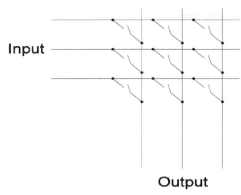

Figure 9.4:
Matrix converter connection circuit

The concept is very simple, consisting of a matrix of switches joining each of the 3 input lines to an output line.

The output voltage waveform is made up of sections of the input as shown in the Figure below. It has been demonstrated that the circuit can operate in all four quadrants with an input line current of any desired power factor.

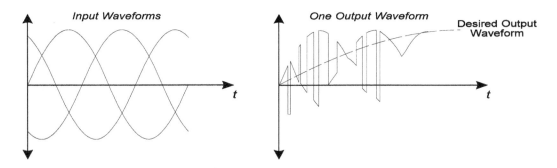

Figure 9.5:
The input and output voltage waveforms of the matrix converter

The main difficulty with the circuit is the requirement that the switches must be able to conduct and block in both directions. Although it is possible to fabricate devices using two power switches and two diodes, this is not an economical solution. There have been attempts to produce a single chip 'universal switch' over the last three years with no commercial success to date.

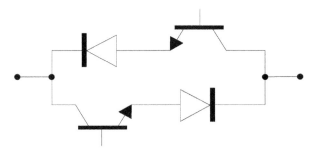

Figure 9.6:
Four-quadrant semiconductor switch

Appendix A

Motor protection – direct temperature sensing

A.1   Introduction

The main requirements for the protection of AC converters and AC induction motors has been covered in considerable detail in Chapter 5: Protection of AC converters and motors. This appendix covers some of the detail of the direct temperature sensing methods of protecting electric motors.

A.2   Microtherm (thermostat)

A *thermostat* is a temperature dependent device that uses a bi-metallic strip to change the position of a pair of contacts at the preset rated response temperature. When the temperature exceeds a preset level, the contacts are used to switch an external control device, such as a relay or contactor. To avoid 'hunting', some sort of hysteresis is usually built into the device to ensure that the set and reset take place at different temperatures.

Microtherms, which are commonly used with electric motors, are miniature precision thermostats, sufficiently small for direct insertion into the windings of a motor or transformer to allow a close thermal association with the winding. The contacts, typically rated at 2.5 amp at 240 volt, are capable of switching a contactor or relay directly. Several microtherms are usually fitted into a motor, each designed to operate at a temperature related to the design temperature of the area in the motor where it is placed. Typical strategic locations are the windings, air ventilation path and bearings. The manufacturers of DC motors tend to prefer Microtherms, while the manufacturers of AC motors tend to prefer thermistors, which are described below.

Microtherms are usually used in groups of two, with one group having a rated reference temperature of 5°C or 10°C lower than the other to provide a temperature pre-warning alarm. The second group is used to trip the motor to prevent damage to the winding insulation. On a motor of significant rating and thermal inertia, the pre-warning alarm would give an operator several minutes to clear the process machine or rectify the overload condition before the overload trip signal occurs.

In DC motors, two groups of microtherms are generally used. The mounting position of the first group is usually at the hottest point of the hottest interpole, usually the one carrying the armature current. This location provides protection for armature current overload. The second group is usually located in the shunt field, providing protection for both the shunt coil temperature and the general temperature within the motor.

In a modern shunt wound DC motor, the working temperature of the shunt winding is very similar to that of the armature. Any loss or restriction of the cooling air, which is difficult to monitor other than by direct measurement, will result in a fast rise in the temperature of the field winding and will be detected by the microtherm.

A.3 Thermistor sensors and thermistor protection relays

A thermistor is a small non-linear resistance sensor, which can be embedded within the insulation of a motor winding, to provide a close thermal association with the winding. It is made from a metal oxide or semiconductor material. The relationship between resistance and temperature is non-linear and the resistance varies strongly with small temperature changes around the set point.

By correct positioning, thermistors can be located close to the thermally critical areas, or hot-spots, of the winding, where they closely track the copper temperature with a certain time lag, depending on the size of the thermistors and how well they are installed in the winding.

Thermistors are most easily inserted into the non-rotating parts of motors, such as the stator winding in an AC motor or the interpole and field windings of a DC motor.

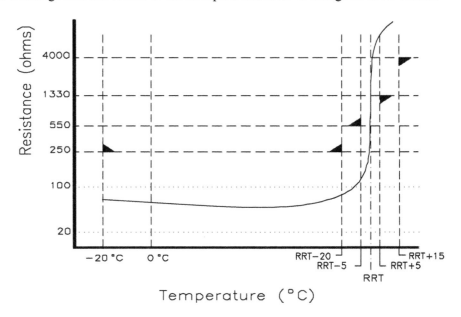

Figure A.1:
Characteristic curve of a PTC thermistor sensor to IEC TC2

RRT = Rated response temperature

IEC specified temperature/resistance limits are clearly marked

The main advantages of thermistors are:

- Their small size allows them to be installed in direct contact with the stator winding.
- Their low thermal inertia gives rapid and accurate response to winding temperature changes.
- They measure temperature directly irrespective of how these temperatures are initiated.
- They can be used to detect overload conditions in motors driven by frequency converters.

The temperature coefficient can be positive (*PTC – positive temperature coefficient*), where the resistance increases with temperature, or negative (*NTC – negative temperature coefficient*), where the resistance decreases with temperature. The type most commonly used in industry is the PTC thermistor, whose typical resistance characteristic is shown in the curve below.

The resistance at normal temperatures is relatively low and remains nearly constant up to the *rated response temperature (RRT)*. As the RRT is approached and exceeded, the gradient of the resistance increases sharply, giving the PTC thermistor a high sensitivity to small changes of temperature. At the set point, a temperature rise of a few degrees results in a large increase in resistance. The resistance is monitored by a thermistor protection relay (TPR) and, when the sharp change in resistance is detected by the thermistor protection relay (TPR), it operates a contact to initiate an alarm or to trip the protected device.

Thermistor protection relays are required to trip reliably when the sensor resistance rises above about 3 kΩ. They will also respond to an open circuit, either in the cable or the thermistor sensor, thus providing fail-safe protection. Modern TPRs are also designed to detect a thermistor sensor short circuit, when sensor resistance falls below about 50 Ω. The specified operating levels are:

- Thermistor over-temperature protection according to IEC
  - Response level = 3300 $\Omega \pm$ 100 Ω

  - Reset level = 1650 $\Omega \pm$ 100 Ω
- Thermistor short-circuit protection according to IEC
  - Response level $\leq$ 15 Ω

In AC variable speed drives, PTC thermistors are commonly used to protect the AC squirrel cage motor fed from inverters. Many modern AC converters have a thermistor protection unit built into the converter, avoiding the requirement for a separate thermistor protection relay.

In DC motors, PTC thermistor sensors are increasingly used instead of microtherms, which are described in the section above.

The rated response temperatures (RRT), which are commonly selected for the various classes of insulation on electric motors, are summarized in the table in Figure A.2.

Insulation class	Rated temp	Alarm temp	Trip temp
Class B	120°C	120°C	130°C
Class F	140°C	140°C	150°C
Class H	165°C	165°C	175°C

Figure A.2:
Typical temperature level settings used on rotating electrical machines

Due to the relatively slow transfer of heat to the sensors through the insulation medium, PTC thermistors do not provide sufficiently fast protection for short circuits in motors or transformers. Also, since they are usually located in the stator windings, they do not provide adequate protection for rotor critical motors or for high inertia starting or stalled rotor conditions. In these cases, to achieve complete protection, it is recommended that PTC thermistors should be used in combination with electronic motor protection relays, which monitor the primary current drawn by the motor.

The application of PTC thermistors as temperature sensors is only effective when:

- The rated response temperature (RRT) of the thermistor is correctly selected for the class of insulation used on the winding.
- The thermistors are correctly located close to the thermally critical areas.
- There is a low thermal resistance between the winding and the PTC thermistor. This depends on the electrical insulation between the winding and the thermistor. Since thermistors need to be isolated from high voltages, it is more difficult to achieve a low heat transfer resistance in HV motors, which have greater insulation thickness.

Several thermistor sensors may be connected in series in a single sensor circuit, provided that the total resistance at ambient temperatures does not exceed 1.5 kΩ. In practice, and as recommended by IEC, up to six thermistor sensors can be connected in series.

For a 3-phase AC motor, two thermistor sensors are usually provided in each of the 3 windings and connected in two series groups of three. One group can be used for alarm and the other group for tripping of the motor. The alarm group is usually selected with a lower *rated response temperature (RRT)*; typically 5°C or 10°C lower than the tripping group. If the operator takes no action, the tripping group is used to trip the motor directly to prevent damage to the winding insulation. In many cases, users choose both groups to have the same RRT. In this case, only one group of thermistors is used (one in each phase) and these are then used for tripping the motor. This provides for one spare thermistor in each phase.

The physical location of the thermistor sensors in an AC motor depends on the construction of the motor, whether it has a cylindrical rotor or salient pole rotor, and several other design and manufacturing variables. In some cases, the optimum location may have to be determined from test experience.

Thermistor protection relays (TPR) are designed for mounting inside a control cubicle or motor control center, usually on standard terminal rails. The Figure A.3 shows a typical connection of two thermistor protection relays, and their associated groups of thermistor sensors.

For alarm and trip control of a 3-phase AC induction motor. The performance of thermistor protection relays can be affected by external electrical interference, where voltages can be induced into the sensor cable. Consequently, cables between the thermistor protection relay and the PTC thermistor sensors should be selected and installed with a view to minimizing the effects of induced noise. Cables should be kept as short as possible and should avoid running close to noisy or high voltage cables over long distances.

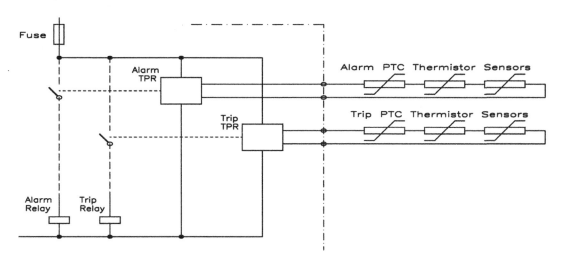

Figure A.3:
Typical connection of thermistor protection relays

During testing, care should be taken not to *megger* across the thermistors as this can damage them. The correct procedure is to connect all the thermistor leads together and to apply the test voltage between them and earth or the phases.

Some practical recommendations for the type of cables that should be used are as follows:

Distances ≤ 20 m   **Standard parallel cable is acceptable**
Distances ≥ 20 m, ≤ 100 m   **Twisted pair cable is necessary**
Distances ≥ 100 m   **Screened twisted pair (STP) cable is necessary**
High level of interference   **Screened twisted pair (STP) cable is necessary**
                                The screen should be earthed at *one end only*

For cable distances to the sensors of greater than 200 meters, the cross-sectional area of the conductors should also be considered. The following are recommended:

Conductor Cross-section	Maximum Length	Type of Cable
0.5mm^2	200m	Screened twisted pair (screen earthed at one end only)
0.75mm^2	300m	Screened twisted pair (screen earthed at one end only)
1.0mm^2	400m	Screened twisted pair (screen earthed at one end only)
1.5mm^2	600m	Screened twisted pair (screen earthed at one end only)
2.5mm^2	1000m	Screened twisted pair (screen earthed at one end only)

Figure A.4:
Recommended cable size to thermistor sensors

A.4   Thermocouple

Thermocouples consist of two lengths of dissimilar metals joined at one end to form a junction. At the other open end, a small voltage is produced which is dependent on the temperature at the junction. This is known as the *peltier effect*. As the temperature changes, the developed thermionic voltage changes to give an indication of the temperature.

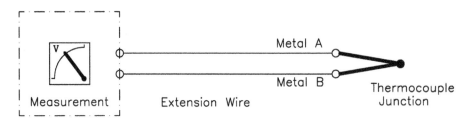

Figure A.5:
Connections between a thermocouple sensor and its controller

There are several national standards, which specify the performance characteristics of thermocouples, such as voltage/temperature, error limits and color-codes for connecting wires. The most commonly used standards are listed below. These standards are generally interchangeable in terms of their relationship between voltage and temperature.

- **ANSI M96.1 – American National Standards Institute (also known as the NBS standard)**
  This is one of the most widely used standards for instrumentation. The ANSI color-code always uses a red negative leg with a different color for the positive leg indicating the type and temperature range of the thermocouple. The overall sheath color is brown.

- **BS 1843 – British Standard**
  This standard uses a blue negative leg color-code, with a different color sheath and positive leg indicating the type and temperature range of the thermocouple.

- **DIN 43714 – German Standard**
  This standard uses a red positive leg color-code, with a different color sheath and negative leg indicating the type and temperature range of the thermocouple.

- **JIS C1610 – Japanese Standard**
  This standard uses a red positive leg and white negative leg color-code, with a different color sheath indicating the type and temperature range of the thermocouple.

- **NF C42-323 – Normes Françaises (French Standard)**
  This standard uses a yellow positive leg color-code, with a different color sheath and negative leg indicating the type and temperature range of the thermocouple.

- **IEC 584 – International Electrotechnical Commission (IEC)**
  This is a new standard that will start to gain acceptance in the future. Hopefully, it will overcome much of the confusion that currently exists with thermocouple color-codes.

Type	Temp range	Metals +ve –ve	NBS colors	BS colors
J-Type	0°C to 750°C	Iron Constantan	White/Red Brown Sheath	Yellow/Blue Black Sheath
K-Type	–200°C to 1250°C	Copper Nickel	Yellow/Red Brown Sheath	Brown/Blue Red Sheath
N-Type	–270°C to 1300°C	Nicrosil Nisil	Orange/Red Brown Sheath	Orange/Blue Orange Sheath
E-Type	–200°C to 900°C	Chromel Constantan	Purple/Red Brown Sheath	Brown/Blue Brown Sheath
T-Type	–200°C to 350°C	Copper Constantan	Blue/Red Brown Sheath	White/Blue Blue Sheath

Figure A.6:
Details of the most common base-metal thermocouples, with the ANSI-NBS and BS 1843 color-codes

As temperature sensors, thermocouples have the following main advantages:

- Robust: very suitable for the industrial environment
- Good accuracy: typically 0.5% per 1°C
- Low cost: consist of a junction of two dissimilar metals
- Self powered: thermal energy is converted into electrical energy
- Wide temp range: types are available for most temperature ranges

The materials used for thermocouple junctions are either base metals or noble metals. base-metal thermocouples are most commonly used in industry because of their lower cost. Thermocouples made from noble metals are more expensive and are used for special applications, where corrosion may be a problem. There are also a number of very high temperature thermocouples, usually made from tungsten.

Conventional copper wire should not be used to connect thermocouples to the temperature controller. This would introduce additional junctions into the circuit and lead to substantial temperature sensing errors. Special thermocouple wire, using the same materials as the thermocouple junction, should be used to connect the junction to the controller. Thermocouple extension wires are usually also color-coded to match the thermocouple colors.

Thermocouple connections are also susceptible to external electrical interference and induced voltages, superimposed onto the junction voltage, will result in measurement errors. Extension wires should not be run along cable routes together with high voltage or high current power cables. Screened extension wires should be used in cases where there is a high level of noise. On industrial sites, it is common practice to run thermocouple extension cables inside galvanized iron conduits, which provide physical protection and shielding against electrical noise.

A.5   Resistance temperature detector (RTD)

Name of Sensor	Metal	Resistance at 0°C
Cu-10	Copper	10 Ω
Pt-100	Platinum	100 Ω
Ni-120	Nickel	120 Ω

Figure A.7:
The most common types of RTD sensors

Resistance temperature detectors (RTDs) monitor temperature by measuring the change of resistance of an accurately calibrated resistive sensor, usually made of copper, platinum or nickel. Tungsten is sometimes used for high temperature applications. RTD sensors can be of the wire wound type, which have a high stability over a period of time, or can be of the metal film types, which are lower cost with faster response but their characteristics can deteriorate over a period of time.

The type of RTD sensor most commonly used in electrical machines comprises a **Pt-100 sensor** element made of platinum, whose resistance is accurately calibrated to 100 Ω at 0°C. The sensor is usually insulated and mounted inside a cylindrical metal tube of dimensions typically 10 mm diameter and 200 mm length.

Since the RTD sensor is physically larger than other types of measuring sensors, it cannot easily be mounted in the windings or bearings of small electric motors. Consequently, RTDs are only used on large machines, where they are installed within the stator slots during manufacture. A slightly different mechanical form is used for mounting in bearing housings. Thermistors or thermocouples are still the most commonly used temperature sensors for electric motors.

An RTD has a linear relationship between resistance and temperature, typically 0.4 Ω/°C for a Pt-100 sensor. A very sensitive measuring instrument, usually based on the Wheatstone bridge, is required to continuously measure the small changes in the resistance of the RTD. These instruments pass a small excitation current through the resistive sensor.

Although the excitation current can cause some problems with self-heating, this is seldom a problem because the currents are small, typically less than 1 mA, and RTDs have a high rate of heat dissipation along the connecting wires and to the measured medium.

Considering the small changes of resistance with temperature, the overall accuracy of the RTD resistance measurement is affected by the series loop resistance of the extension wire between the measuring instrument and the Pt-100 sensor. This is dependent on the cross-sectional area of the wires and the distance between the RTD sensor and the measuring instrument. This has led to the development of 3-wire RTDs, where a third identical extension wire is connected between the instrument and the sensor. The purpose of the third wire is to provide the measuring instrument with a means of measuring the wire loop resistance to the RTD sensor. To improve accuracy, this is subtracted from the total measured resistance. At the RTD sensor, the third wire is simply connected to one of the legs of the sensor as shown in Figure A.8.

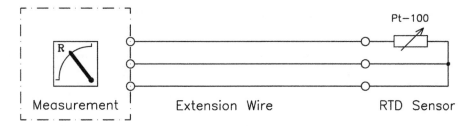

Figure A.8:
Connections for a 3-wire resistance temperature detector

In a similar way to thermistors and thermocouples, the RTD connections are also susceptible to external electrical interference and induced voltages, which can lead to measurement errors. Similar precautions need to be taken with the cable route selection and screening. RTDs have become very popular in industry because they provide low cost, high accuracy temperature measurement with a relatively fast thermal response.

Pt-100: Resistances for temperatures 0°C to +299°C : DIN 43760

°C	+ 0°C	+ 1°C	+ 2°C	+ 3°C	+ 4°C	+ 5°C	+ 6°C	+ 7°C	+ 8°C	+ 9°C
0	100.00	100.39	100.78	101.17	101.56	101.95	102.34	102.73	103.12	103.51
+10	103.90	104.29	104.68	105.07	105.46	105.85	106.24	106.63	107.02	107.40
+20	107.79	108.18	108.57	108.96	109.35	109.73	110.12	110.51	110.90	111.28
+30	111.67	112.06	112.45	112.83	113.22	113.61	113.99	114.38	114.77	115.15
+40	115.54	115.93	116.31	116.70	117.08	117.47	117.85	118.24	118.63	119.01
+50	119.40	119.78	120.17	120.50	120.94	121.32	121.70	122.09	122.47	122.86
+60	123.24	123.62	124.01	124.39	124.77	125.16	125.54	125.92	126.31	126.69
+70	127.07	127.46	127.84	128.22	128.60	128.99	129.37	129.75	130.13	130.51
+80	130.89	131.28	131.66	132.04	132.42	132.80	133.18	133.56	133.94	134.32
+90	134.70	135.08	135.46	135.84	136.22	136.60	136.98	137.36	137.74	138.12
+100	138.50	138.88	139.26	139.64	140.02	140.40	140.78	141.15	141.53	141.91
+110	142.29	142.67	143.04	143.42	143.80	144.18	144.55	144.93	145.31	145.69
+120	146.06	146.44	146.82	147.19	147.57	147.95	148.32	148.70	149.07	149.45
+130	149.83	150.20	150.58	150.95	151.33	151.70	152.08	152.45	152.83	153.20
+140	153.58	153.95	154.33	154.70	155.07	155.45	155.82	156.20	156.57	156.94
+150	157.32	157.69	158.06	158.44	158.81	159.18	159.56	159.93	160.30	160.67
+160	161.05	161.42	161.79	162.16	162.53	162.91	163.28	163.65	164.02	164.39
+170	164.76	165.13	165.50	165.88	166.25	166.62	166.99	167.36	167.73	168.10
+180	168.47	168.84	169.21	169.58	169.95	170.32	170.69	171.05	171.42	171.79
+190	172.16	172.53	172.90	173.27	173.64	174.00	174.37	174.74	175.11	175.48
+200	175.84	176.21	176.58	176.95	177.31	177.68	178.05	178.41	178.78	179.15
+210	179.51	179.88	180.25	180.61	180.98	181.34	181.71	182.08	182.44	182.81
+220	183.17	183.54	183.90	184.27	184.63	185.00	185.36	185.73	186.09	186.46
+230	186.82	187.18	187.55	187.91	188.27	188.64	189.00	189.37	189.73	190.09
+240	190.45	190.82	191.18	191.54	191.91	192.27	192.63	192.99	193.35	193.72
+250	194.08	194.44	194.80	195.16	195.53	195.89	196.25	196.61	196.97	197.33
+260	197.69	198.05	198.41	198.77	199.13	199.49	199.85	200.21	200.57	200.93
+270	201.29	201.65	202.01	202.37	202.73	203.09	203.45	203.81	204.16	204.52
+280	204.88	205.24	205.60	205.96	206.31	206.67	207.03	207.39	207.74	208.10
+290	208.46	208.82	209.17	209.53	209.89	210.24	210.60	210.96	211.31	211.67

Figure A.9:
Pt-100 sensor – variation of resistance with temperature over range 0°C to +299°C

Pt-100: Resistances for temperatures 0°C to –219°C : DIN 43760

°C	–0°C	–1°C	–2°C	–3°C	–4°C	–5°C	–6°C	–7°C	–8°C	–9°C
0	100.00	99.61	99.21	98.82	98.43	98.03	97.64	97.25	96.86	96.46
–10	96.07	95.68	95.28	94.89	94.49	94.10	93.71	93.31	92.92	92.52
–20	92.13	91.73	91.34	90.94	90.55	90.15	89.75	89.36	88.96	88.57
–30	88.17	87.77	87.38	86.98	86.59	86.19	85.79	85.40	85.00	84.61
–40	84.21	83.81	83.42	83.02	82.63	82.23	81.83	81.44	81.04	80.65
–50	80.25	79.85	79.46	79.06	78.66	78.26	77.87	77.47	77.07	76.68
–60	76.28	75.88	75.48	75.08	74.68	74.28	73.89	73.49	73.09	72.69
–70	72.29	71.89	71.49	71.09	70.69	70.28	69.88	69.48	69.08	68.68
–80	68.28	67.88	67.47	67.07	66.67	66.26	65.86	65.46	65.06	64.65
–90	64.25	63.84	63.44	63.03	62.63	62.22	61.82	61.41	61.01	60.60
–100	60.20	59.79	59.39	58.98	58.57	58.16	57.76	57.35	56.94	56.54
–110	56.13	55.72	55.31	54.90	54.49	54.08	53.68	53.27	52.86	52.45
–120	52.04	51.63	51.22	50.81	50.40	49.98	49.57	49.16	48.75	48.34
–130	47.93	47.52	47.10	46.69	46.28	45.86	45.45	45.04	44.63	44.21
–140	43.80	43.38	42.97	42.55	42.14	41.72	41.31	40.89	40.48	40.06
–150	39.65	39.23	38.82	38.40	37.98	37.56	37.15	36.73	36.31	35.90
–160	35.48	35.06	34.64	34.22	33.80	33.38	32.96	32.54	32.12	31.70
–170	31.28	30.86	30.43	30.01	29.59	29.16	28.74	28.32	27.90	27.47
–180	27.05	26.62	26.20	25.77	25.34	24.91	24.49	24.06	23.63	23.21
–190	22.78	22.35	21.93	21.50	21.08	20.65	20.23	19.80	19.38	18.95
–200	18.53	18.11	17.70	17.28	16.86	16.44	16.03	15.61	15.19	14.78
–210	14.36	13.96	13.57	13.17	12.78	12.38	11.99	11.59	11.20	10.80

Figure A.10:
Pt-100 sensor – variation of resistance with temperature over range 0°C to –219°C

Appendix B

Current measurement transducers

B.1   Current shunt

A current shunt and amplifier is the simplest form of current feedback. This well established technique is illustrated in Figure B.1 the DC bus current is passed through a link of pre-calibrated resistance and the voltage across it is measured. The voltage is directly proportional to the current passing through the shunt.

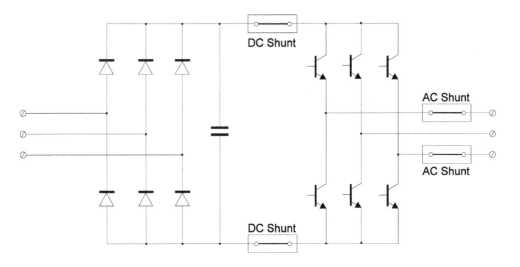

Figure B.1:
Shunt voltages to earth on an AC converter

The main problem with this device is one of electrical isolation. To measure the DC bus current in a PWM AC converter, the shunt must be located in the positive or negative leg of the DC bus, and will therefore be at some voltage above (or below) earth potential as shown in the Figure B.1. This problem of isolation also occurs when measuring currents in the output phases to the motor.

The simplest way to overcome this problem is to reference the control circuit to the shunt potential, which may be around 300 V above earth. While this approach was adopted in many early AC VSDs, it is no longer considered acceptable as it poses interface and safety problems when control devices are connected to the drive.

Another approach is to galvanically isolate the shunt circuit from the rest of the control circuit with an *isolation amplifier*. This can be achieved with discrete circuitry incorporating either opto-couplers or signal transformers. However, it adds significantly to the complexity and cost of the current feedback circuitry.

Current shunts are seldom used in digital VSDs. Hall effect sensors are far more common.

B.2 Principle of the Hall effect sensor

The fundamental principle of the *Hall effect sensor* is shown in Figure B.2. The current flowing through a semi-conductor material establishes a magnetic field in a plane perpendicular to the current, which forces the moving carriers to crowd to one side of the conductor. As a result of this crowding, a *Hall voltage* will develop perpendicular to both the current and the magnetic field.

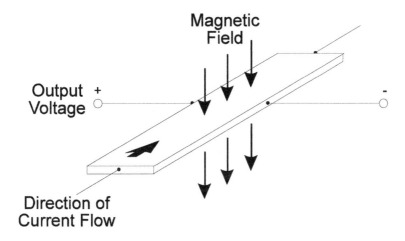

Figure B.2:
Principle of the Hall effect current sensor

B.2.1 Open-loop Hall effect sensor

If a Hall effect semi-conductor integrated circuit is placed in the air-gap of a toroidal magnetic core enclosing a current carrying conductor, the output voltage of the Hall device is proportional to the current in the main conductor. This configuration is shown in the Figure B.3 below. Thus a current measurement signal is produced which is inherently isolated from the primary current being measured and with a small number of components. This device also has a relatively high bandwidth of around 10 kHz.

While this method is simple and cheap to manufacture, the performance of this Hall device is variable and each combination of core and device must be carefully trimmed. This is usually done at the desired current trip point for the drive so as to achieve maximum accuracy during fault conditions. Measurement errors will then be present at normal operating currents and will be greatest at zero current. In practice, this is not usually a major problem in general purpose AC drives, as the current signal is mainly used for current limit control and power device protection.

However, because of these inherent problems, the open-loop Hall effect sensor is not commonly used on larger VSDs, but is quite common for small VSDs.

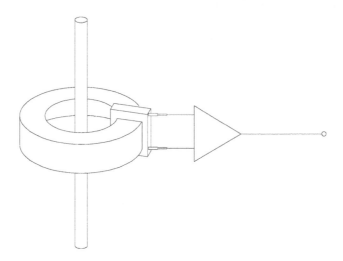

Figure B.3:
Principle of the open-loop Hall effect current sensor

B.2.2 Closed-loop Hall effect sensor

A method of overcoming the accuracy and repeatability problems of the open-loop sensor is to include a feedback loop to null the flux in the magnetic core. This is the basic principle of the well-known *LEM* (manufacturer's name) Hall effect current sensing modules, which are widely used in most modern VSDs. The configuration of this device is shown in Figure B.4 below.

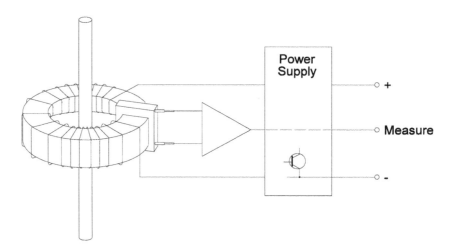

Figure B.4:
Principle of the LEM Hall effect current sensor

In the closed-loop module, the secondary winding has many more turns than the primary winding, which carries the primary current being measured. An amplifier is used to drive current into this secondary winding so that the net magnetic field in the core is zero. At this point, the secondary current, I_S is given by:

$$I_S = I_P \times \frac{\text{No. of Primary Turns}}{\text{No. of Secondary Turns}}$$

The main advantage of the closed-loop method is high accuracy and repeatability over a large signal range. Bandwidth is still good (around 10 kHz), although the performance of the amplifier is critical in achieving high bandwidth. These models also have significant power supply requirements, as the secondary winding current will be related to the maximum primary current times the turns ratio as shown in the equation above. For example, a LEM module with a 1000:1 turns ratio, will draw 100 mA if the primary current to be measured is 100 A.

In addition to the high current requirement, the LEM usually requires a DC power supply of about 15 V DC in order to achieve a high rate of change of current in the secondary winding. This high di/dt is also necessary to maintain high bandwidth.

While closed-loop devices are generally considered superior due to their excellent accuracy, repeatability and good bandwidth, they tend to be used mainly in larger drives. Smaller VSDs (below 4 kW) cannot justify their high cost and power supply requirements, so the open-loop Hall sensor is commonly used at this power level.

Appendix C

Speed measurement transducers

C.1   Analog speed transducers

A tachometer generator or *tacho-generator* (abbreviation: *tacho*) is a small electromagnetic generator that is usually fitted to the non-drive end (NDE) shaft of an electric motor. Tacho-generators can be either of the flange-mounted, solid-shaft type or the hollow-shaft type.

Since *tachos* are usually small in size relative to the motors to which they are attached, the flange-mounted type is susceptible to damage due to excessive axial forces. Consequently, special care needs to be taken with the coupling between the tacho and motor shaft to avoid problems with alignment and the difficult fitting and removal operations associated with maintenance. Misalignment can also result in a low frequency ripple that is difficult to filter out. The commonly used 'bellows' type of coupling is designed to absorb axial forces and vibration, which reduce the life of the brushes, and also allows for minor misalignment. External magnetic fields can also affect the output of a tacho-generator.

Hollow-shaft tachos overcome many of the mounting difficulties of the flange-mounted tachos. The following are the two most common methods of mounting hollow-shaft tachos:

- **Hollow-shaft tacho with separate rotor and stator parts**
  The rotor is mounted directly onto a stub-shaft at the NDE end of the rotating machine, using either a keyway or a friction press-fit. The stub-shaft diameter is typically in the range 8 mm to 16 mm. The stator is fixed onto a flange on the bearing housing of the machine. No bearings are supplied with this type of hollow-shaft tacho. However, care must be taken to ensure that the tacho rotor is concentric with the tacho stator. The brush holders and cable connections are mounted on the stator frame.

- **Integral hollow-shaft tacho with bearings**
  These units are complete with bearings and suitable for direct mounting onto a stub-shaft at the NDE end of the rotating machine. To prevent the stator from

rotating, a mounting bracket is used to lock it to the bearing housing of the machine.

Tacho-generators are available with DC or AC outputs as follows:

- ### The DC tacho-generator
  This is a permanent magnet DC generator comprising a permanent magnet stator with a wound rotor armature with commutator and brushes. The output of the DC tacho is a DC voltage, whose magnitude is directly proportional to the rotational speed of the motor and whose polarity depends on the direction of rotation. Usually, the output voltage is arranged to be positive for clockwise rotation and negative for counter-clockwise rotation.

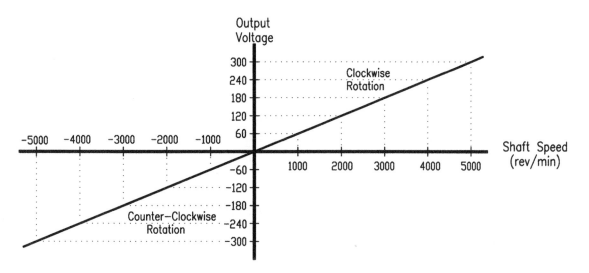

Figure C.1:
The output voltage of a 60 volt/1000 rpm bi-directional DC tacho

A good quality DC tacho-generator should meet the following requirements:

- Output voltage precisely proportional to the speed of rotation in both directions

- Negligible difference between the output voltage for specific speeds in both the clockwise and counter-clockwise directions of rotation

- Stability of the output voltage in relation to temperature and time

- Low ripple in output DC voltage

- Small size with low moment of inertia of rotating parts

- Sturdy electrical and mechanical construction for high reliability

- Class of enclosure protection to suit industrial environment

- Low requirement for maintenance

- Adequate speed range for most common applications

When used for variable speed drive applications, tacho-generators are usually rated for an output voltage of 60 volts per 1000 rpm (speed constant = 0.06 V/rpm) with a maximum speed of up to 5000 rpm. This results in a maximum output voltage of up to 300 volts. However, other output voltages and speed ranges are also available to suit specific requirements. The class of protection of the tacho enclosure is usually specified to be the same as the machine to which it is attached, e.g. IP54, IP55 or IP56.

- **The AC tacho-generator**
  This is very similar to a DC tacho, except that the output is an alternating AC voltage, whose magnitude and frequency is directly proportional to the rotational speed. The AC voltage output is usually rectified and converted into a DC voltage in the control system. Consequently, the AC Tacho is unsuitable for bi-directional applications because the resulting voltage magnitude after rectification is unipolar. Also, a high ripple content in the voltage signal at low speed makes the AC Tacho unsuitable for low speed applications.

  AC tacho-generators are seldom used with variable speed drives.

C.2 Digital rotary speed and position transducers

A digital rotary encoder is an electromechanical transducer, which converts rotary speed or position into a series of digital electronic pulses. Rotary encoders are commonly used for the following applications:

- Feedback of rotary speed from electric motors for variable speed drive control
- Determining the angular position of rotating machines for synchronizing movements
- Tracking the position of robots, stackers, reclaimers and other automated machines
- Monitoring the position of products on a conveyor
- Measuring the length of materials, which are fed from rolls, for cutting to size

The two main types of digital encoder are as follows:

- **Incremental rotary encoder**
  Generates a series of square voltage pulses as the shaft rotates. External electronic circuitry is required to determine the speed of rotation, direction, angular position or length of material fed off a roll by counting the number of pulses or measuring the rate at which they are transmitted by the encoder.

- **Absolute rotary encoder**
  Generates a parallel code, comprising 4, 8, 12 or 16 bits, which represents the angular position of the shaft. The Gray code is most commonly used for absolute encoders. Occasionally, the binary or BCD code are also used.

C.2.1 Incremental rotary encoder

Incremental rotary encoders are the type most commonly used for VSDs and other industrial positioning and synchronizing systems. The series of voltage pulses, which represent the rotational speed, are generated by the interaction of light from a light source passing through a pattern of lines printed onto a transparent rotating disk.

The disk is usually made of a laminated plastic material to provide low weight and inertia with a high resistance to mechanical shock. Some manufacturers use disks made of glass, but these tend to be more fragile and heavier than the plastic types. Accurately spaced lines, with spaces of identical width (space/width ratio = 1), are printed in a radial pattern around the outer edges of the disk. Encoders with line counts of up to about 10 000 lines are available.

A beam of light, produced by an LED or miniature lamp with a focussing lens, is passed through the graduated transparent disk and detected by a sensor on the other side. Except for some very low rate incremental encoders, a stationary 'mask', with a line pattern identical to that on the disk, is required to alternately block or pass the light beam as the disk rotates. The intensity of the resulting light is sensed by a photo-cell (transistor), which switches on when the light beam is passed and switches off when the light beam is interrupted. This technique is known as the moire fringe principle.

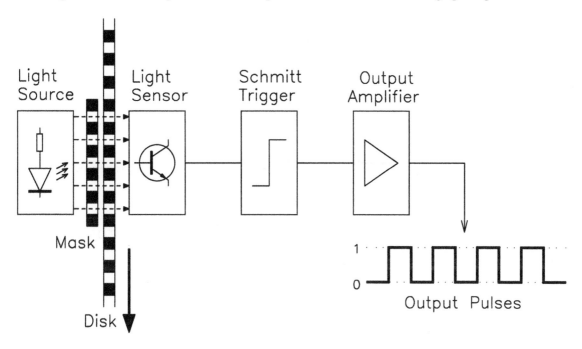

Figure C.2:
The principle of an incremental encoder

The stationary mask usually has 3 separate fields, each with the same line spacing as the rotating disk. Three photo-cells detect the presence/absence of light via these three masks and pass the information to three output Channels. The first mask/sensor is for Channel-A. The second mask/sensor is for Channel-B, which lags Channel-A by 90°. The third mask/sensor is for the reference Channel-O, which provides one output pulse per revolution.

A *Schmitt Trigger* squares off (digitizes) the electronic output of the photo-sensor before passing the signal through an amplifier to produce a train of squared output pulses with a constant amplitude. Amplitude is not important because the receiver is interested only in the pulse rate and their phase relationship to each other.

The rate of the digitized output pulses from an incremental encoder depends on:

- Resolution, the number of lines printed around the perimeter of the disk
  This is normally specified as *lines per revolution* or *pulses per revolution (ppr)*. The resolution is sometimes also given as the angular distance between two consecutive lines on the perimeter of the rotating disk.

- Rotational Speed, the rotational speed of the shaft of the machine

To measure rotational speed, an incremental rotary encoder requires only one track of lines around the perimeter of the rotating disk, making it much simpler internally when compared to an absolute encoder. The train of squared voltages is available from an output Channel usually designated as *Channel-A*. An additional inverted signal (complement of Channel-A), is usually available for signal noise immunity (see Section C.2.3 on Output interfaces). The actual speed is measured by external electronic equipment that uses up/down counters to determine speed and position. The external electronic equipment usually also provides the power supply to the encoder.

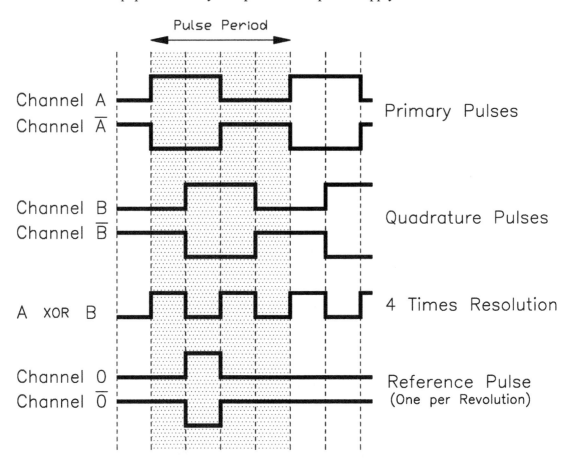

Figure C.3:
Squared output pulses from a typical incremental rotary encoder

To measure direction, a second train of pulses is required from a similar *Channel-B*, which is offset by 90° lagging relative to Channel-A. Channel-B is also known as the quadrature Channel. The quadrature Channel-B uses the same track of lines on the disk perimeter, but uses a separate mask grid with a 90° offset. An inverted Channel-B signal is also usually provided.

The counters in the external electronic equipment monitor the relationship between Channel-A and Channel-B to determine shaft speed and direction. This feature also provides the ability to multiply the incremental value by four, using a built-in exclusive-OR (XOR) gate, to increase the resolution.

For example, a incremental encoder with a line count of 10 000 can provide an output of 40 000 steps per revolution.

- Clockwise direction is determined when Channel-A leads Channel-B by 90°
- Counter-clockwise direction is determined when Channel-B leads Channel-A by 90°

The third *Channel-O* provides one output pulse per shaft revolution. This signal can be used to synchronize the serial pulse train with a known mechanical position. In this case, care should be taken during fitting to ensure that the position of the Channel-O pulse on the disk is mechanically aligned with the required external reference point.

The following are common resolution values (ppr) for incremental rotary encoders:

2	4	8	16	32	36	48	50	60	90	100
120	125	128	150	180	200	250	256	300	320	360
375	400	420	480	500	512	600	625	720	800	900
1000	1024	1080	1100	1250	1270	1500	1600	1728	1750	1800
2000	2048	2400	2500	2540	2700	3000	3600	3750	4000	4096
4200	4500	5000	6000	6400	7200	8192	9000	10000		

The physical arrangement of the components inside an incremental rotary encoder depend on the manufacturer. A cross-section of a typical construction is shown below.

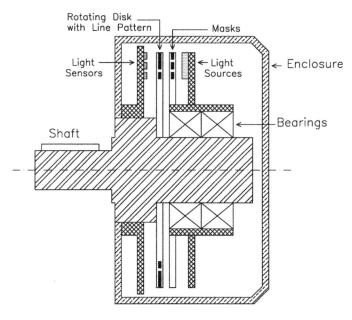

Figure C.4:
Physical arrangement of a typical incremental rotary encoder

C.2.2   Absolute rotary encoder

Single-turn absolute rotary encoders generate a parallel digital output, which represents the angular position of the shaft over one revolution. Each angular position is represented by a code, comprising several digital bits. The rotating optical disk has multiple tracks, one for each bit of the output code. Even if the power fails, once restored, the position of the shaft is accurately known without any resetting routine, as would be required for incremental encoders.

One revolution (360°) is divided into a specified number of positions. The number of possible positions (*position resolution*) depends on the number of tracks printed onto the disk (number of bits in the code) as shown in the table on the next page.

No. of tracks (bits in code)	Resolution (No. of positions)	Resolution (deg per position)	Resolution (deg/min/sec)
1	2	180.00°	180° 00' 00'
2	4	90.00°	90° 00' 00'
3	8	45.00°	45° 00' 00'
4	16	22.50°	22° 30' 00'
5	32	11.25°	11° 15' 00'
6	64	5.63°	5° 37' 30'
7	128	2.81°	2° 48' 45'
8	256	1.41°	1° 24' 23'
10	1024	0.35°	0° 21' 06'
12	4096	0.088°	0° 05' 16'
16	65 536	0.0055°	0° 00' 20'

Figure C.5:
The relationship between the number of tracks (bits) on the disk and the resolution of an absolute rotary encoder

Absolute rotary encoders are available with outputs using one of the following codes:

- The binary code
- The binary coded decimal (BCD) code
- The Gray code
- The Gray excess code

Figure C.6 shows the 6-bit coded output signal for absolute encoders with 6 tracks using the binary, BCD and Gray codes for a resolution of 36 positions (10°) over one revolution. This sequence can be reduced or extended to match any required resolution.

The **binary absolute encoder** uses a binary number (counting system to the base 2) to represent the position number of the shaft. This is best illustrated by way of an example.

Example:

What is the parallel 8-bit output code of an 8 track (256 position) single-turn binary absolute encoder, which is stationary at a 270° shaft position?

- An 8 track absolute encoder has 256 (2^8) possible positions, numbered 0 to 255

- Position at $270°$ $\Rightarrow$ position-191

- The decimal position is converted to its equivalent in binary

- Position-191 $\Rightarrow$ binary position code = 10111111

The procedure to convert any decimal number (base 10) to its equivalent binary number (base 2) is to repeatedly divide the decimal number by 2 until the quotient becomes zero.

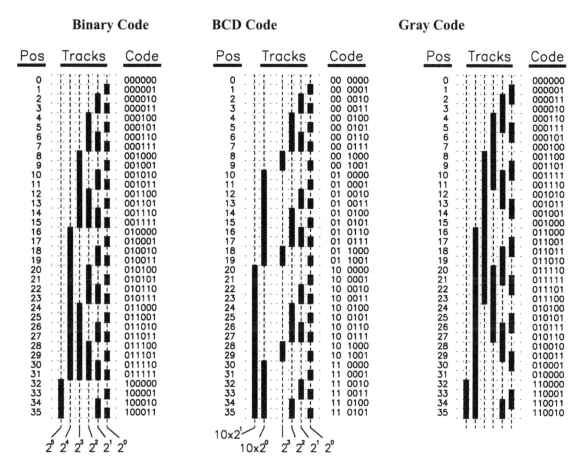

Figure C.6:
6-bit binary, BCD and Gray codes for a resolution of 36 positions

The remainders at each stage are the individual bits of the binary number:

- 191/2 = 95, remainder 1 (least significant Bit or LSB)

- 95/2 = 47, remainder 1

- 47/2 = 23, remainder 1

- 23/2 = 11, remainder 1

- $11/2 = 5$, remainder 1

- $5/2 = 2$, remainder 1

- $2/2 = 1$, remainder 0

- $1/2 = 0$, remainder 1 (most significant bit or MSB)

The remainder $= 10111111$, which is the 8-bit code for $270°$

The procedure to convert the binary position number back to its equivalent decimal position number is much simpler. Each bit of the binary number represents the multiplication factor of the exponents of base 2 as illustrated below.

- Binary $= 10111111$

- Decimal $= 1 \times 2^7 + 0 \times 2^6 + 1 \times 2^5 + 1 \times 2^4 + 1 \times 2^3 + 1 \times 2^2 + 1 \times 2^1 + 1 \times 2^0 = 191$

Although the binary code is used extensively in digital electronic equipment, it has a major drawback for rotary encoders. As the shaft position changes from one angular position to the next, several bits in the code are required to change state simultaneously. For example, in Figure C.6, five bits change state between position 15 to 16. If there are slight differences between the switching times of the photo-cells reading the different tracks, the resulting output code would represent a false position.

The **BCD (binary coded decimal) code** is an extension of the binary code. The four *least significant bits (LSB)* of the BCD code follow the binary code, while the higher order bits in the BCD code represent the decimal values as follows:

bit 1: 2^0.(LSB)
bit 2: 2^1
bit 3: 2^2
bit 4: 2^3
bit 5: 10×2^0
bit 6: 10×2^1
bit 7: 10×2^2
bit 8: 10×2^3.(MSB)
etc

The BCD code is the most convenient code to display shaft position on a set of 7-segment LED modules. The 7-segment LED modules are usually arranged in a decimal format with *units* on the right-hand side, then *tens*, then *hundreds*, then *thousands*, etc. BCD encoders suffer from the same disadvantages as the binary encoders.

The Gray code is the preferred code for absolute encoders, particularly when used for position control. The Gray code avoids the reading error problems associated with binary and BCD encoders because only one bit changes between any two neighboring positions (see Figure C.6). Reading error due to a slight misalignment of the photo-cells is reduced to only 1 step.

However, when the required resolution (no. of positions) on an absolute rotary encoder is a value less than 2^x (where x = no. of tracks), several bits need to change at the zero position of the disk and the Gray encoder would have the same disadvantages as the binary encoder.

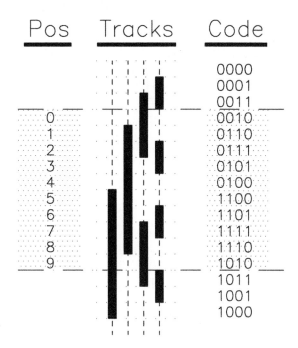

Figure C.7:
An example of the Gray excess code: 10-excess-3 Gray code

The Gray excess code has been developed to overcome this problem and consists of only a *middle* portion of the Gray code. For example, with a 4 track absolute encoder, a resolution of 2^4 (16 positions) is possible. If only 10 positions are required per revolution and the Gray code was truncated after position 9, three bits would have to change to return the code to position 0, as shown in Figure C.7. If the first 3 positions are omitted from the pattern on the rotating disk and the following 10 positions used, then a change of only 1 bit can be maintained for all positions of the disk. The Gray excess code for this example would then be referred to as a *10-excess-3 Gray code*.

Another common example is an absolute encoder, which uses the Gray excess code, is one with a position resolution of 360 positions per revolution, or 1° per position. In this case, a disk of at least 9 tracks is required to provide a 9-bit Gray code (2^9 = 512), resulting in a *360-excess-76 Gray code*.

Some industrial applications require an angular position resolution that is greater than ONE revolution. In these cases, **multi-turn absolute rotary encoders** can be used to distinguish between several revolutions. Additional graduated disks are provided with mechanical reduction gearing between them.

The most common multi-turn absolute encoders use a parallel 24 bit Gray code, comprising a main disk with 12 tracks (resolution = 2^{12} or 4096) and up to 3 smaller supplementary disks, each with 4 tracks and geared at a ratio of 16:1.

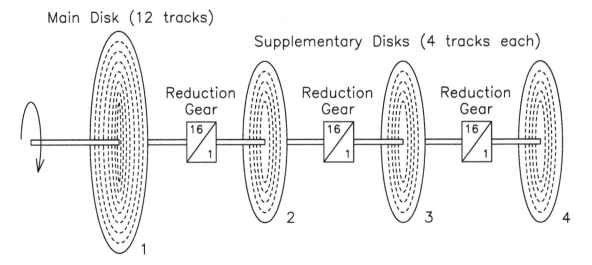

Figure C.8:
Example of multi-turn absolute encoder with 4 disks and gearing

With this arrangement, the following position and turns resolution can be achieved:

- 12 track main disk, resolution of 2^{12} (4096 positions)
- 1×4 track supplementary disk, resolution of 2^4 (16 revolutions)
- 2×4 track supplementary disks, resolution of $2^4 \times 2^4$ (256 revolutions)
- 3×4 track supplementary disks, resolution of $2^4 \times 2^4 \times 2^4$ (4096 revolutions)

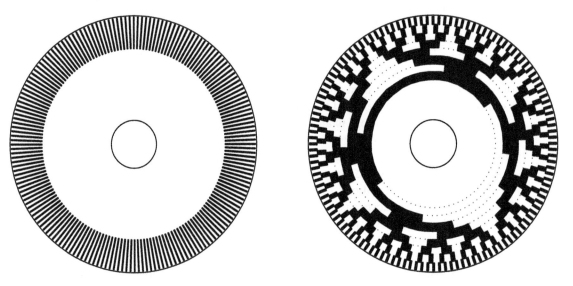

Figure C.9:
Examples of disk patterns for incremental and absolute rotary encoders

C.2.3   Output interface connections

To cater for the various types of electronic circuits available, rotary digital encoders are manufactured with several output options. The maximum output frequency is typically 100 kHz, but can be higher with some units for particular applications.

The type of connection between a rotary digital encoder on a machine, and the electronic decoder in the control equipment some distance away, is determined mainly by the distance between the two. Series volt drop, shunt capacitance and induced electrical noise all affect the quality of the signals.

Several output interface connections are possible between encoders and decoders:

- NPN current sink interface (unbalanced – single wire) are used mainly for inputs to control equipment such as PLC counters
- PNP current source interface (unbalanced – single wire) are used mainly for inputs to control equipment such as PLC counters
- Push-Pull – NPN or PNP interface are used mainly for inputs to control equipment such as PLC counters
- Line driver interface (balanced differential – two wire EIA-422) are used mainly for speed feedback from digital encoders to variable speed drive controllers

For short distances between the encoder and electronic decoder, up to approximately 10 m, and where externally induced noise is not a major problem, a simple type of encoder line driver will provide adequate results. In these cases, an unbalanced (single-wire) type of connection is adequate for each Channel. Power for the encoder is normally provided from the decoder end and the output is connected in an unbalanced configuration.

Voltage and current levels depend on the type of decoder, which can be 5 V (TTL-standard), 12 V or 24 V. The latter two voltages are often used with PLCs.

Typical circuit configurations are as follows:

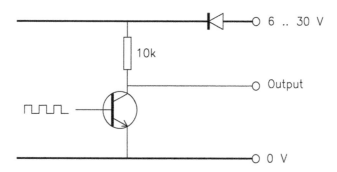

Figure C.10:
NPN current sink interface (unbalanced – single wire) with a pull-up resistor to the +volt supply line and switching to 0 volt line

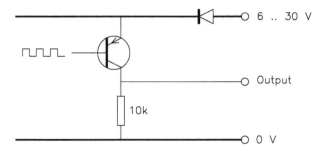

Figure C.11:
PNP current source interface (unbalanced – single wire) with a pull-down resistor to the 0 volt line and switching to +volt line

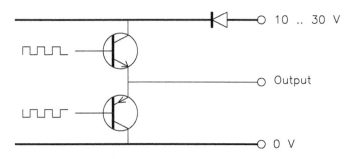

Figure C.12:
Push-pull unbalanced line driver can be used

For longer distances between the encoder and electronic decoder, of up to ±100 m, and where externally induced noise is more of a problem, a balanced differential line driver is necessary to achieve acceptable reliability. With rotary encoders this is the most commonly used circuit. In these cases, a two-wire interface (per channel), based on the well-known EIA-422 (RS-422) interface standard is commonly used for all channels. Each channel produces two voltages, being the signal voltages (A, B & O) and their complement (A, B & O). Power for the encoder is usually provided from the decoder end. For this type of connection, terminating resistors are necessary at the decoder end to avoid reflections.

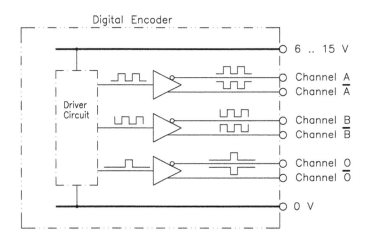

Figure C.13:
EIA-422 balanced differential line driver with remote power supply

For even longer distances between the encoder and electronic decoder, of up to ±1000 m, and where externally induced noise and power supply voltage drop are a problem, a balanced differential line driver (EIA-422) with a **local power supply source** is necessary to achieve acceptable reliability. In this case, terminating resistors are also necessary at the decoder end to avoid reflections.

On modern absolute encoders, to reduce the number of connecting wires and to improve the performance in areas of high electrical noise, additional internal electronics is used to convert the parallel code (up to 24 bits) to an equivalent serial data code for transmission, usually using synchronous communication over 2 or 4 wires.

Appendix D

International and national standards

D.1 Introduction

International trade, and the acceptability of imported and exported products, is dependent on the existence of internationally acceptable standards and norms. This appendix summarizes the various international and national standards, which are applicable to rotating electric machines and variable speed drives.

D.2 International standards

Standardization is coordinated internationally by the **ISO – International Standards Organization**, which draws its members from all countries of the world. ISO concentrates on coordinating standards internationally. ISO has published almost 9000 standards, covering a broad range of subjects. To facilitate international trade, there is a general movement in most countries towards bringing their own standards into harmony with the International standards. Standardization is achieved through the participation of interested parties, usually from national standards organizations, interested academics, scientists and manufacturers. Reaching consensus on standards internationally is a tedious process and it is coordinated through:

- **TC**: Technical committees
- **SC**: Technical sub-committees
- **WG**: Working groups

Standardization of electrotechnical matters is coordinated internationally by the **International Electrotechnical Commission (IEC)**. In a similar way to ISO, standardization is achieved through the participation and consensus of interested parties, usually from national standards organizations, interested academics, scientists and manufacturers. As in ISO, standardization is coordinated through:

- **TC**: Technical committees
- **SC**: Technical sub-committees

- **WG**: Working groups

D.3   European standards

To a greater or lesser degree, all European countries have developed their own national standards. Since 1945, the major countries of Europe have been actively working towards a common market and, during this period, there has been a great expansion of inter-European trade. Initially this common market was called the European Economic Community (EEC), then the European Community (EC) and most recently the European union (EU). To overcome the technical differences between the products manufactured in European countries, several committees have been established to harmonize standards. In some cases, new standards known as European norms (EN) have been introduced.

The following committees have been established to coordinate European standardization:

- **CEN**: Comité European de Normalisation (Committee for European standardisation)
- **CENELEC**: Comité European de Normalisation – Electrotechnique (Committee for European Electrotechnical Standardization)
- **CCITT**: Comité Consultatief International Telegraphique et Telephonique (consultative committee for international telegraph and telephone)
- **ECISS**: European Committee for Iron and Steel Standards

Because of the economic power of Europe, these 'Norms' will inevitably have relevance to non-European countries, which seek to trade with Europe. Many of these ENs will become *de facto* international standards or be embodied in ISO or IEC standards.

D.4   United States of America standards

Economically and militarily, the USA has been the most powerful and influential country in the world for at least the past 50 years. The sheer size of its economy and the large number of high caliber technical experts working there has provided the USA with an unusually high capacity for technical innovation. Many well-known standards, which have become *de facto* international standards, have been developed in the USA. Most electrical engineers are familiar with the USA standards, developed by organizations such as ANSI, FCC, IEEE, NEMA, MIL, EIA/TIA and UL. Many of these have been formally embodied in ISO and IEC standards. Also, because of the large defense industry in USA, a wide range of military (MIL) standards have been developed for difficult environments. These are usually significantly tighter than their civilian counterparts.

In spite of this economic and technical power, standardization in the USA is largely out of step with the rest of the world. This is mainly a result of the different units of measure used in the USA compared to the rest of the world. Almost all countries outside the USA, such as Western Europe, Eastern Europe, Japan, Canada, Australia & New Zealand, Africa, South America, China, Taiwan and the fast growing SE Asian countries have based their industries on the metric system of measurement and have actively tried to harmonize their standards along the lines of ISO and IEC. On the other hand, the USA is only partially metricized and has largely kept the old English system of measurement, which is based on feet, inches, pounds, gallons, etc. In general, the USA has largely been reluctant to move in the direction of harmonizing their standards with ISO and IEC. This continues to create difficulties with international free trade.

D.5   National standards

All industrialized countries have their own national standards bodies, which establish and coordinate national standards. These standards organizations are usually members of ISO and IEC.

To facilitate international trade, national standards are increasingly 'harmonized' with the relevant ISO and IEC standards. Any differences that still remain between the national and international standards are usually associated with some special local conditions and established local practice.

The national standards bodies of the larger industrialized countries of the world are listed in the table shown in Figure D.1.

D.6   South African standards

In order to harmonize with the rest of the world, *SA Bureau of Standards* has resolved to adopt international standards wherever possible, provided that they are appropriate. This has changed the emphasis away from writing our own standards to looking critically at those from other organizations, such as ISO, IEC and CENELEC and participating in the continual review and updating of those standards.

Country	Initials	Name of Standards Organisation
Australia	AS	Standards Australia
Belgium	IBN	Institut Belge de Normalisation
Canada	CSA	Canadian Standards Association
CIS (ex USSR)	GOST	Gosudarstvenne Komitet Standartov
Denmark	DS	Dansk Standardisieringsraad
	DEMKO	Danmarks Elektrise Material Kontrol
European	CEN	Comite European de Normalisation
	CENELEC	Comite European de Normalisation – Electrotechnique
	CCITT	Comite Consultatief International Telegraphique et Telephonique
Finland	SFS	Suomen Finland Standardisoimisliitto
France	AFNOR	Association Française de Normalisation
	UTE	Union Technique de l'Electricité
Germany	DIN	Deutches Institut für Normierung
	VDE	Verband Deutscher Elektrotechniker
International	IEC	International Electrotechnical Commission
	ISO	International Organisation for Standards
	ITU	International Telecommunications Union
Italy	CEI	Comitato Electrotechnico Italiano
Japan	JIS	Japanese Industrial Standard
Netherlands	NNI	Nederlands Normalisatie Instituut
New Zealand	NZS	Standards Association of New Zealand

Norway	NSF	Norges Standardisieringsforbund
	DNV	Det Norsk Veritas
Poland	PRS	Polish Register of Ships
Saudi Arabia	SASO	Saudi Arabian Standards Association
South Africa	SABS	South African Bureau of Standards
Spain	UNE	Una Norma Española
Sweden	SIS	Standardisieringskommissionen I Sverige
Switzerland	SEV	Schweizerischer Electrotechnischer Verein
United Kingdom	BSI	British Standards Institution
United States of America	ANSI	American National Standards Institute
	FCC	Federal Communications Commission
	IEEE	Institute of Electrical and Electronic Engineers
	NEMA	National Electrical Manufacturers Association
	UL	Underwriters Laboratories

Figure D.1:
The names and initials of the major standards organizations

Appendix E

Glossary of common terms used with AC variable speed drives

A

AC	alternating current, an electrical transmission system where the voltage and the associated current alternatively adopt a positive and negative polarity, typically at 50 cycles per sec
A/D converter	device used to convert analog signals into the digital format
Algorithm	a procedure or set of steps which are used, usually in a computer program or processor, to solve a problem
Analog	a continuously varying waveform that may represent a voltage, current or any other continuously changing value
ANSI	American National Standards Institute, which is the main standards organization in the USA
ASCII	American standard code for information interchange, a 7-bit code, commonly used in data communications, for encoding alphabetic and numeric characters – originally specified by ANSI, it is now also specified as International Standards ISO-646 and CCITT alphabet #5
ASIC	application specific integrated circuit
Asynchronous	a data communications technique where the group of bits, representing a character, can be transmitted at an arbitrary time with a variable 'idle' time between characters – when a character is transmitted, the group of bits is preceded by a start-bit, which synchronizes the receiver, and is terminated by a stop-bit – close synchronization is necessary during transmission of the bit-group

AS	Australian standards of the Standards Association of Australia (SAA), the main standards organization in Australia
Attenuation	the decrease in the magnitude of a signal, sound or voltage down a transmission path – usually measured in decibels or dB

B

Bandwidth	range of frequencies in Hz, being the difference between the lowest and highest frequency, which can be transferred along a transmission medium without significant attenuation
Band pass filter	a filter that permits a limited range of frequencies to pass through. The frequencies outside this range are sharply attenuated
BCD	binary coded decimal, a code that represents each character in a decimal number with a 4-bit code
Bit	binary digit, represented by either a logic condition 1 or 0
BJT	bipolar junction transistor, (also see GTR), 3-terminal power transistor rated for high currents between the emitter and collector terminals, usually comprising a double or triple Darlington connection, which can be turned on and off by means of a current applied to the base terminal
Baud	unit of measurement for signal speed, which is based on events per second (usually bits per second) – if each event has more than one bit associated with it, then the baud rate and bits per second are not equal – *Baud* is derived from the name of the famous French telegrapher, Maurice Emile Baudot
Bps	bits per second, the speed at which bits are transferred
BRC	block redundancy check, an error checking technique in data communications
Bridge	an interface device, used to connect two similar networks, which enables the network to be extended – it is often called a 'repeater'
BS	British standards of the British Standards Institution (BSI), the main standards organization in the United Kingdom
Buffer	a temporary storage location for data (FIFO) in the receiving end device – compensates for the difference in the data flow rates between the transmitting device and the receiving device
Byte	a group of 8 bits, which are usually the code for a character

C

Capacitance	the property of an AC electrical circuit to store electrical energy when a voltage is applied –it is normally associated with two plates, where the capacitance is proportional to the area of the plates and inversely proportional to the distance between them – when an alternating voltage is applied, the resulting current will be such that the current peak will

lead the voltage peak by 90° – units of capacitance are measured in Farads

CAD	computer aided design, a computer based design program.
CAE	computer aided engineering, a computer based engineering program.
CAM	computer aided manufacturing, a computer based manufacturing program.
CENELEC	Comité European de Normalisation Electrotechnique, whose standards are often applied by European countries
Character	refers to any alphabetic letter, numeral, punctuation mark or other symbol, which may be used in the transfer of data
Checksum	the result of binary addition of bits, representing characters in a data message, used for error detection
CNC	computer numerical control, a computer controlled machine
CPU	central processing unit, the intelligent core of a digital device
CRC	cyclic redundancy check, a very effective error checking technique in data communications, using an error message 2 bytes long (16 bits) – two versions are commonly used, being CRC-16 and CRC-CCITT
CSA	Canadian Standards Association, the main standards organization in Canada
CSI	current source inverter, a configuration of frequency converter where rate of change of current (di/dt), behind the inverter, is limited by a large choke connected in series in the DC link

D

Darlington connection	a 'cascade' connection of transistors which increases the amplification factor of a power transistor module, in order to reduce the base current required from the control circuit – double or triple Darlington connections are commonly used in power converters
D/A	device used to convert digital signals into the analog format
dB	decibel – a measure of attenuation in a signal, based on the logarithmic ratio of the signal magnitude V_1 and V_2 at the two ends – the ratio is expressed as, $dB = 20\log_{10}V_1/V_2$
DC	direct current, an electrical transmission system where voltage and current retain a fixed polarity (positive or negative)
DCE	data circuit-terminating equipment, a name from the EIA/TIA standards that applies to a particular configuration of comms port
DCS	distributed control system, an industrial control computer usually used for process control applications
DEC	decimal

Digital	a type of signal that has two or more definite states – a binary digital signal is a special case, which has only two states represented by logic 1 and 0
DIN	Deutches Institut für Normierung, the main standards organization in Germany
DIP	dual in-line package, usually a group of switches used on a PCB
DOL	direct-on-line, a method of starting AC induction motors by switching them directly to the power source via a contactor

E

EIA	Electronic Industries Association, a standards organization in USA that defines the technical details of interface connections in data communications – it recently changed its name to TIA, Telecommunications Industries Association
EMI	electro-magnetic interference, the type of electrical interference induced by electro-magnetic fields in the vicinity of a cable
EPROM	erasable programmable read only memory, a non-volatile memory, commonly used with microprocessors to store program data – data in the memory can be erased, usually by ultraviolet light; data is not lost when the power is removed
EEPROM	electrically erasable programmable read only memory, a non-volatile memory, commonly used with microprocessors to store program data – data in the memory can be erased and updated electronically; data is not lost when the power is removed

F

Farad	unit of measurement of capacitance in the metric system
FCT	field controlled thyristor, a thyristor that can be turned on and off by a voltage applied to the gate
FDM	frequency division multiplexing, a technique that divides the available frequency bandwidth into separate narrower bands to permit several simultaneous communication channels
Fieldbus	a general term given to any standard system which defines the physical and software requirements for connecting the field devices and process controllers together in a control system
FIP	Factory information protocol, the name given to a fieldbus standard, which includes the protocol – it was developed by French companies
FET	field effect transistor, a transistor that can be turned on and off by means of a voltage applied to the gate terminal
FSK	frequency shift keying, a technique for transferring data bits by shifting between two or more particular frequencies

G

G	giga, metric system prefix, $\times 10^9$
GTO	gate turnoff thyristor, a thyristor that can be turned on and off by means of a current applied to the gate
GTR	giant transistor, also called a bipolar junction transistor (BJT), which is a power transistor rated for high currents – it can be turned on and off by means of a current applied to the base

H

Henry	unit of measurement of inductance in the metric system
HEX	hexadecimal, a counting system to base 16
Hz	Hertz, unit of measurement of frequency in the metric system, where 1 Hertz = 1 cycle per second

I

IC	integrated circuit, an encapsulated electronic circuit, containing miniaturized electronic components, that is designed to perform in a particular way
IEC	International Electrotechnical Commission, an international standards organization, which specializes in electrical standards
IEAust	Institute of Engineers Australia, a professional institute for engineers in Australia
IEE	Institute of Electrical Engineers, a professional institute for electrical engineers in UK
IEEE	Institute of Electrical and Electronic Engineers, a professional institute for electronic and electrical engineers in USA
IGBT	integrated gate bipolar transistor, a voltage controlled electronic switching device, similar to a MOSFET
Impedance	a combination of resistance R and reactance X, measured in ohms, that provides an opposition to the flow of current in an electrical circuit – units are measured in ohms
Inductance	the property of an electrical circuit that opposes a current flow, causing the current peak to lag behind the voltage peak by $90°$ – units are measured in henrys
Interface	a common electrical boundary between two separate devices, over which data or other electrical signals can pass between them
I/O	inputs and outputs, the connections into and out of a control device such as a PLC, DCS, RTU, etc
ISA	Instrument Society of America

ISO	International Standards Organization, an organization that coordinates standards internationally
ISP	interoperable systems project, the name given to a project that is aimed at reaching finality on a standard fieldbus system, being a compromise between Profibus and IEC-SP50, where data communicating devices from several manufacturers can communicate *interoperably* on the same network

K

k	kilo, metric system prefix, 10^3
kVA	kilovolt-amperes, measurement of, volt $\times$ amp $\times 10^3$
kW	kilowatt, measurement of, watt $\times 10^3$

L

LAN	local area network, a data communications system, connecting several communicating nodes, where the communication cables are shared – it is usually restricted to a small geographic area
LCD	liquid crystal display, a visual display system for operators, using liquid crystals
LCI	load commutated inverter, an inverter in which the thyristors are turned off by the electrical behavior of the load device
LED	light emitting diode, a diode that emits light when current is passing through it

M

m	meter, the unit of length in the metric system
M	mega, metric system prefix, $\times 10^6$
mho	unit of measurement of conductance in the metric system
min	minute, measurement of time = 60 sec
MODEM	MOdulator-DEModulator, a device that converts digital voltage data to frequencies suitable for transmission over an analog communications system, such as a telephone or radio channel
MOS	metal oxide semiconductor, a semiconductor device, using a specific type of construction
MOSFET	metal oxide semiconductor field effect transistor, an FET using the MOS construction – it is a voltage controlled electronic switching device
MOV	metal oxide varistor, a non-linear semiconductor device used for over-voltage protection

μ**P**	microprocessor, an 'intelligent', miniature, encapsulated processor, used for controlling digital circuit
MTBF	mean time between failures, a statistical measure of the average period between failures of any component of a device
MTTR	mean time to repair, a statistical measure of the average downtime, after a device has failed, before it can effectively be put back into service
Multidrop network	a communication network, where three or more similar devices are connected, or 'multidropped', onto one network

N

NVPROM	non-volatile programmable read only memory, a non-volatile microprocessor memory, with relatively fast access, commonly used with microprocessors to store program data – the data in the memory can be erased and updated electronically and is not lost when the power is removed
Nm	Newton meters, the measurement of torque in the metric system
NEMA	National Electrical Manufacturing Association, an association that publishes standards for electric power, construction and testing codes in USA
Network	a communications system where several devices share the same communications channel
Node	the point where a communicating device is connected into a network
Noise	the undesirable electrical signals that are induced into a communications network from other neighboring electrical equipment carrying high voltages or high currents – noise often results in distortion of the signal or data errors

O

Ohm	unit of measurement of resistance & impedance in metric system
Optical isolation	a technique for galvanically isolating two electronic circuits by means of a light path – the signal is transferred over the isolating barrier by using a light-emitting source, such as an LED, and a light sensitive receiver, such as a transistor

P

PAM	pulse amplitude modulation, a modulation technique commonly used in the inverter of older generations of AC variable speed drives to control the amplitude of the output AC voltage
Parity	an error checking system on the bits that make up a character – even parity: parity bit is set so that the number of 1 bits are even – odd parity: parity bit is set so that the number of 1 bits are odd

PC	personal computer, a microprocessor based computer designed for personal, office and industrial use, such as a PC-AT with 286, 386 or 486 microprocessor
PC	programmable controller, a computer for use in industrial control and which can be programmed by the user to carry out a particular control sequence
PCB	printed circuit board, a flat piece of insulation material that supports a number of electronic components and onto which an electrical circuit has been etched by photographic means
PLC	programmable logic controller, a computer, originally designed for digital sequence control in industrial applications – modern PLCs can also do calculations and monitor analog inputs or control analog outputs – they can be programmed by the user to carry out a particular control sequence
Port	the terminals on a communicating device, used for input and output of digital or analog signals
Profibus	process fieldbus, the name given to a standardized field network system, including the protocol, developed by mainly German and other European companies
PROM	programmable read only memory, a non-volatile microprocessor memory, with relatively slow access, which usually holds a program to control the sequence of the control device – the program stored in the memory can be programmed *once* by the user but cannot be changed thereafter; it is not lost during power down
Protocol	a formal set of rules that specify all the software, flow control, error detection and timing requirements for the exchange of messages between two communicating devices on a network
PWM	pulse width modulation, a modulation technique used in the Inverter of many modern AC variable speed drives to control the amplitude of the output AC voltage and to improve the shape of the current waveform

Q

QA	quality assurance, a management and documentation procedure aimed at the close supervision of all aspects of design, manufacture, testing and installation of any device or plant
QC	quality control, the supervision procedures of quality assurance

R

RAM	random access memory, a read/write volatile memory with fast access to/from a microprocessor – it is used for holding temporary data during calculation and/or implementation – data in the RAM is lost during power down, unless battery backup is provided

Reactance	the opposition to the flow of current when an alternating voltage is applied to an electrical circuit, due to the inductance of the circuit ($X_r = j\omega L$ ohms) or the capacitance of the circuit ($X_c = 1/j\omega C$ ohms)
Resistance	the opposition to the flow of current when a voltage is applied to an electrical circuit, measured in ohms – resistance = voltage/current
RF	radio frequency, which refers to high frequency waveforms above the audible range
RFI	radio frequency interference, noise whose frequencies are in the radio frequency range
ROM	read only memory, a non-volatile microprocessor memory, with relatively slow access, which holds a program to control the sequence of a control device – the program is stored in the ROM during manufacture, cannot be changed by the user, and is not lost during power down
RS-232	now also called EIA-232 or TIA-232 – it is a standard that defines the electrical and mechanical details of the physical interface between two devices, designated DTE and DCE, employing serial binary data interchange, over an unbalanced copper wire interface
RS-422	now also called EIA-422 or TIA-422 – it is a standard that defines the electrical details of the physical interface between two or more devices, employing serial binary data interchange, over a balanced differential copper wire interface
RS-485	now also called EIA-485 or TIA-485 – it is a standard which defines the electrical details of the physical interface between two or more devices, employing serial binary data interchange, over a balanced differential copper wire interface – it defines transmitters with tri-state facilities suitable for multidropping, allowing up to 32 transmitters and 32 receivers on a network
RTD	resistance temperature device, a type of thermometer sensor where the temperature is directly proportional to the resistance
RTU	remote terminal unit, an input/output terminal device that can be mounted remotely (far away) from a programmable controller – communication between them can take place via a wire, fiber, radio, modem, carrier or any other suitable medium

S

SAA	Standards Association of Australia, the main standards organization in Australia
SCIM	squirrel cage induction motor, a design of induction motor that is very commonly used in industry
SCR	silicon controlled rectifier, an alternative name for a thyristor
SEC	State Energy Commission, name of the corporation that operates the power supply systems in Australia

sec	second, a measurement of time
Serial comms	a data communications transmission mode, where the bits are transferred sequentially on a single data channel
SI	Systemes Internationale, the international system of units – SI was adopted in 1969 as ISO Recommendation R1000
SMT	surface mount technology, a technique for mounting the electronic components on the surface of a PC board
STP	shielded twisted air, usually comprising an insulated pair of copper wires, twisted together to reduce the effect of induced noise – a conductive shield around the twisted pair, earthed at one end only, provides screening against capacitively coupled noise – STP can be used for data transmission with speeds of up to 20 Mbps

T

TDM	time division multiplexing, a technique which divides the available transmission time into narrow time blocks to permit several communication channels to use the same link – a demultiplexer is used at the receiving end to separate the different messages
Thyristor	a thyristor is a semiconductor 'switch' with 3 terminals, being the anode, cathode and gate – with a forward voltage across the anode–cathode terminals and a current applied to the gate (G), current will flow into the anode (A) and out of the cathode (K) – the thyristor is switched off when the anode current is reduced $\leq$ zero
TEFC	totally enclosed fan cooled, an electric motor cooling method
TO/L	thermal overload (relay), a device that monitors current and uses a bimetal strip to simulate the thermal characteristics of an electrical load, such as an electric motor
TPR	thermistor protection relay, an electronic device that monitors the resistance of a thermistor and closes a relay contact when the resistance exceeds a preset level
TPU	thermistor protection unit, another name for a thermistor protection relay (see above).
TTL	transistor–transistor logic, which uses positive DC logic with +5 volt = logic **1** and 0 volt = logic **0**

U

UTP	unshielded twisted pair, usually comprising an insulated pair of copper wires, twisted together to reduce the effect of induced noise – no conductive screen is provided and it can be used for data transmission with speeds of up to 20 Mbps

UL	Underwriters Laboratory Inc, a public testing Institute in USA

V

VDE	Verband Deutcher Elektrotechniker, German Standards by DIN, which are generally in harmony with the equivalent IEC standards
VSD	variable speed drive, controls the speed of a motor, or the mechanical device coupled to it, either mechanically or via an electronic control circuit
VSI	voltage source inverter, a configuration of frequency converter where rate of change of voltage (dv/dt) behind the inverter is limited by a large capacitor connected in parallel across the DC link
VVVF	variable voltage variable frequency (converter), where both the voltage and the frequency output of the converter are controlled to control the speed and limit the flux in the AC motor

W

WAN	wide area network, a data communications system, connecting several intelligent devices, where the communication channels are shared – it is usually used over a large geographic area

Index

Printed and bound by CPI Group (UK) Ltd, Croydon, CR0 4YY

03/10/2024

01040334-0020